Foreword

I always hoped that Vic Priel would write a book. Contact with him over many years, when I and a few other devoted pioneers were striving to "put maintenance on the map," convinced me that here was an individual with the dedication and determination to accumulate the data and experience which would enable him to give expression to his strong beliefs on this subject.

His work over the last decade has exposed him to a wide range of industries in varying states of development in many countries, in a capacity which has enabled him to experience at first hand all the pressures of industrial life in the realm of maintenance work, from the shop floor to managerial responsibility.

This knowledge has been concentrated in this book and it is safe to say either to one who is new in the field of maintenance responsibility or to one who is seeking to improve the effectiveness of an established organisation that the answers to their questions lie between these covers. It will be a matter, after reading his observations on the strengths and weaknesses of a number of alternative techniques, of selecting those which will best serve the short- and long-term needs of the reader's particular circumstances and construct the organisation which will be most effective.

That this can now be done with greater certainty as to the results is due to this book which will be equally appropriate on the bookshelf of every supervisor and manager and in training establishments which deal with this important subject.

G. G. Corder, A.M.B.I.M.

Formerly Assistant Chief Engineer, Kodak Ltd.
Author of *Maintenance—techniques and outlook*,
published by the British Productivity Council

SYSTEMATIC MAINTENANCE ORGANISATION

V. Z. PRIEL

B.Sc. (Mech. Eng.), M.Sc. (Syst. Anal.),
Diploma Eng. Prod. and Management,
C.Eng., M.I.Prod.E., A.M.B.I.M.

MACDONALD & EVANS LTD
8 John Street, London WC1N 2HY

1974

First published April 1974

ISBN: 0 7121 1926 4

To Son-hee
without whose patience
and encouragement this book
would not have been completed

Printed in Great Britain by
Richard Clay (The Chaucer Press), Ltd,
Bungay, Suffolk

Preface

Management of the maintenance function is an empirical science. It embraces the techniques of industrial management whose purpose is to ensure the best utilisation of human, technical and financial resources. By employing the most appropriate management techniques the maintenance executive strives to reach the ultimate objective, namely, *the perfect balance between the input cost of the resources and the performance of good maintenance practice.*

Scientific know-how is now available which, if well applied, could allow the maintenance manager to increase the efficiency of his function. His task, however, is not an easy one. Good maintenance practice does not depend wholly on the degree to which he applies management techniques. Success often depends on factors outside his control, such as technological change or turnover of personnel. His work is also affected by the economic fortunes of his company and sometimes even by its geographical location. But most of all, the service he is able to provide depends on the attitude of management towards his function and on the degree of co-operation that he can achieve with other functions inside the company. Management cannot discharge its responsibilities towards maintenance by saying, "Given that amount of money you are expected to produce such and such results." The proper performance of maintenance requires management's active and constant support.

Maintenance is a conglomerate of activities which strongly affects the surrounding functions and which in turn affect the operation and performance of maintenance. It is a closed cycle which sometimes becomes a vicious circle.

A coherent maintenance function requires specific relationships with accounting, personnel, engineering and production management. The successful maintenance manager is one who is able to involve all these functions in the planning of his objectives and in the performance of established procedures. Thus maintenance becomes an integral part of the planning, operation and control of the company as a whole. How to achieve this is a major objective of this text.

This book aims to show the administrative/managerial techniques necessary for setting up a maintenance function able to cope with the existing needs of the company. At the same time, it also attempts to present the complicated relationships existing between the essential elements of these techniques, thus enabling the practising manager to adapt his ideas to variable conditions and combine them into an effective system.

The analytical, do-it-yourself approach offered here should enable the maintenance manager to appraise his work and make the necessary adjustments towards increasing the effectiveness of his department. The systematic framework of this book presents the separate parts of maintenance in such a way as to allow them to be assembled like a jigsaw puzzle to produce, finally, an integrated system. The techniques offered can either be adopted directly or adapted to particular circumstances.

Special emphasis in this book has been placed on making the text readable and easily understood. Most of the material is as applicable to small plants as to large ones. Much of the argument, however, is particularly valid for small to medium plants where maintenance is seldom treated with respect.

The main features of the book lie in its illustrations, and the text is intended to support the development of a systematic sequence and the reasoning behind it at every stage. Since the material has been collected from experience as a consultant over the years it is hoped that part of it will be familiar to my former colleagues who are practising in this field.

Maintenance know-how has made great progress during the last thirty years, Coverage in the technical press has led to widespread acceptance of new principles and at the same time has provided the world with a vast store of technical information.

Furthermore, top-grade exponents and practitioners in this field are now able to present and discuss their viewpoints at the international conferences which are held regularly in Chicago and London, and occasionally in Tokyo.

Since the first draft of this book was written a number of techniques have been formalised into separate branches of specialisation. The most prominent among them are the reliability concept and maintainability—which already rate engineering titles in the U.S.A.—and logical fault diagnosis. Other techniques from the industrial management field have also been successfully applied in maintenance. Several of these techniques are in fact offshoots of operational research (O.R.) such as network analysis (PERT–CPM), the queueing theory and statistical stock control. Since the present volume concentrates on organisational systems no attempt has been made to cover the above-mentioned subjects, important though they may be. They are only briefly referred to and some of the basic books and articles which discuss them are listed in the bibliography, cross-references to which are made by the small numbers that occur in the text.

We who work in this field hope that our profession will achieve the status for which the early pioneers fought. To earn the recognition of industry we must base our work on a well-defined body of knowledge. The author hopes, therefore, that in this endeavour the present book will make a contribution.

V.Z.P.

January 1974

Contents

List of Illustrations

List of Tables

Chapter 1

Why is maintenance so special?

Any person in charge of maintenance whether in a small or large company feels that his is a unique field. It presents as great a challenge to a person's engineering abilities as it does to his management know-how. It requires as much tact and tolerance as it requires patience and persistence—in fact, it requires a superman. Therefore we may well ask, Why is maintenance so hard to understand, so hard to manage and so hard to prescribe? In this chapter we will attempt to answer these questions in detail.

THE IMPORTANCE OF MAINTENANCE

The case for proper maintenance has been frequently made and does not need to be repeated here. By now we expect industry to be thoroughly maintenance-conscious. However, let us review in brief the factors that make modern industry so dependent on maintenance.

The profusion of equipment

Mankind consumes today an unprecedented amount of goods and makes use of innumerable services. Whatever is consumed must first be produced, whether it is a bar of soap, a can of food or a typewriter; the chain of equipment used in their production is interminable. Because we can easily visualise the bustling activity involved in producing goods, we tend to associate maintenance with manufacturing and forget that cars, buses, railways, ships and planes also have to be continually maintained. In mining and other heavy industries maintenance is of prime importance. Large blocks of flats as well as modern industrial buildings represent vast investments in materials and service equipment which must be maintained over long periods if they are to produce an adequate return. Similarly, in the hotel business or in hospital management, there is a great deal of complex equipment which has to be maintained in such a manner that the public is never aware that maintenance is necessary.

It has become part of our modern way of life to expect our mechanical slaves to function perfectly at all times and if they fail us we unhesitatingly scrap them. The more sophisticated the equipment becomes, the better we expect it to perform, and with the continuation of this trend we become more dependent on maintenance.

The increased complexity of processes

The creation of new tools and mechanical devices has been rapidly accelerating in recent years. In the not-too-distant past tools were mainly an extension of man's natural capacity, while present-day tools harness natural forces, complement and magnify the power of the senses, increase the speed of work, enlarge the capacity of output and ensure accuracy of operation. To accomplish these things the complexity of machines has increased enormously.

Take the paper industry, for instance, and its allied branches of printing and packaging. We observe how manual printing, folding and binding have been superseded by machines which perform at staggering speeds, while at the same time a whole range of hydraulic, pneumatic and electronic devices ensure accuracy of operation at every stage. It is these devices which, in conjunction with management techniques such as "quality control," ensure the high degree of interchangeability and the uniformity of products that we have come to expect.

To assist in the maintenance of such complex equipment new approaches have been developed, among them the "reliability concept" and "logical fault-finding." No less important in the field of maintenance are novel techniques for detecting and predicting malfunctions.

Together with the increasing complexity and rising costs of such equipment there has been an upsurge in the intensity of its utilisation. Our expectations for higher and better output have accentuated the need for better maintenance. In one field, however, that of computers, where most people would expect to find a higher degree of complexity, there are signs of a reversal in this trend. The latest "third-generation hardware," using solid-state circuitry, is smaller, simpler, more efficient in operation and easier to maintain than the earlier types—manufacturers call this phenomenon "increased reliability." Car manufacturers, perhaps as a direct result of difficulties encountered by their customers in maintaining over-elaborate machines, are trying to introduce drastic changes which will reduce the need for servicing.

The cost of production

Cost consciousness in industry is sharpened by the highly competitive nature of modern commerce. Nowadays with new materials appearing almost daily, it is not only a case of one washing powder competing against all others, but disposable clothing, if it is ever produced economically, may decrease both the sale of cloth and that of washing powder.

Then there is the competition of one country with another, where the imported product competes against the locally made one. The markets in some cases involve not only countries but often whole continents with enormous volumes of sales. To capture a share of these markets products have to be competitively produced and without good maintenance this is impossible.

In this constant ebb and flow of commercial pressures there is both an urgent need for prompt deliveries as well as sensitivity to the profit margin. Both aspects are greatly affected by the right application of maintenance. Maximum output and minimum production costs coupled with prompt delivery schedules are essential to success, and this is where maintenance can play an important role. In the field of services, catering and utilities good maintenance makes the difference between profit and loss in a very obvious way. Road transport, for instance, is competing against rail, and shipping against transport by air. Similarly, hotels compete with each other and against motels or caravan-users. The customer goes where he gets the best service at the lowest price.

In this commercial struggle the high cost of investment is an important factor. An adequate life-span and a high resale value for equipment is ensured by proper maintenance, which therefore becomes a major contributing factor in maximising returns on capital. In developing countries, where capital is scarce, the preservation of investment is of prime importance.

Technical progress and obsolescence

When technical progress is as rapid as it is at present, we must be very selective in making technical decisions. Problems relating to the overhauling of used machines, the purchase of new machines and the stocking of spares constantly confront us.

Maintaining the existing plant poses problems in the stocking of spares, especially with model changes, when the spare parts may change both their shape and their catalogue number. Another problem is in the "repairability" of items installed—frequently, the fact that some units are supposed to be discarded after a specified time is usually discovered at the most inconvenient time.

Progress inevitably leads to obsolescence. Processes become obsolete as a result of technological change or perhaps because new materials have made them unnecessary. Our maintenance system must take these factors into account. Maintaining machines beyond their useful life is wasteful. We realise, therefore, that technological change poses problems to the maintenance staff, apart from the financial headaches that the chairman of the board has to endure. Furthermore, maintenance has to be able to adapt to change. Change creates specific needs, none of which are easy to satisfy, such as:

(*a*) the need to keep up to date;
(*b*) the need to use new tools, materials, instruments and techniques;
(*c*) the need to understand and assess the maintenance requirements of new processes;
(*d*) the need to change outdated routines.

A maintenance system can operate efficiently only if it can absorb changes in the technical field as well as shifts in emphasis from one kind of service to another. More often than not it is taken for granted that members of the maintenance staff will keep abreast of developments—they are expected to respond instinctively to change. When new techniques are introduced craftsmen are expected to teach themselves how to deal with the new equipment in the normal course of their duties.

It is bad enough that most maintenance personnel have for a great number of years acquired their skills while at work, and mostly by trial and error. It is more regrettable that most of them usually keep working in the same old way, oblivious of time and change. But can we assume that this haphazard approach will satisfy our future needs? Since technical progress is racing ahead, can we afford to tolerate this situation? The answer must be no, since it is essential that maintenance staff be well informed and up to date; therefore the reading of books and periodicals should be encouraged. This view is, unfortunately, not widely accepted. After all, the argument goes, how could the use of complex equipment ever have gained acceptance were it not for our ability to find ways to cope with it?

Technological progress coupled with a competitive drive also focuses attention on the need for maintaining machinery after the buyer installs it and starts using it. However, no item of machinery can be sold with a guarantee of trouble-free performance and manufacturers often assume the existence of first-class maintenance services. The ability of such services to deal with whatever comes their way becomes absolutely essential.

Thus, we see that there is a great number of reasons why maintenance is fast gaining the recognition that it deserves. The archaic approach of building a machine, a product or a factory and then worrying about keeping it in good order is giving way to forethought and planning.

The conflict between the need to minimise operating costs and the need to have money set aside for maintenance expenditure arises mainly where such planning is absent from the start. A similar clash occurs between the need for equipment availability and the need for time off for servicing. There are further disadvantages when no allowances have been made for natural deterioration and obsolescence.

In fact, the principles of good management entail the consideration of a complete *chain of events*. Those who are engaged in manufacturing cannot limit their attention to the making and selling but must also think of the purchasing, warehousing and shipping of their product as well as providing for the maintenance of their manufacturing facilities. These duties form an integral part of the process in the same way as transportation is part and parcel of producing goods and services.

From a maintenance point of view a car requires servicing and an occasional change of tyres, while a child requires clothing, washing and an occasional dose of medicine. In other words, maintenance is an intrinsic part of existence-and-operation and is an axiom of ownership. Instead of giving it grudging acceptance, we should face it squarely and "tool up for it."

It is surprising, therefore, that maintenance is often ignored, pushed aside, and its importance denied. However, this attitude is now slowly disappearing. With the phenomenal rise of professional management we can expect better implementation and an improved understanding in the future.

Acceptance of the need for effective maintenance is nevertheless clouded by misunderstandings. The problem is inherent in the functions that we are called upon to perform. It is somewhat akin to the services of a dentist. If we visit him regularly costs are stable and the treatment may not be too painful. If we ignore recurrent symptoms and delay in visiting him the costs can rise steeply and the treatment often becomes very painful indeed. If we need him too frequently we may blame *him* for being unable to take care of *our* teeth. If we are not aware of any cavities we are only too eager to skip our next six-monthly visit. This is the see-saw reasoning that is widespread in respect of maintenance.

The following section will explain the characteristic features of maintenance, including some features that are not too well known, and at the same time it may help to clear up some of the prevailing misunderstandings.

CHARACTERISTICS OF MAINTENANCE

There are certain aspects of maintenance that set it apart from other industrial functions. There is a sharp contrast between the way in which we work towards the goals of production and the means through which we strive towards the objectives of maintenance. The former are familiar to everybody, they are easy to explain and the results are evident. Placing emphasis on producing more, obtaining better quality and even the reduction of waste during production are obvious and definable objectives.

Amid this activity, a production-orientated executive must accept the maintenance function whose aims do not conform to the familiar frame of reference and which may even appear to conflict with his normal operations. While all his efforts are concentrated on producing more, reducing costs and purchasing only what is needed, maintenance service seems to distract him from his goals and reduce his achievements. Thus it is not surprising that a recurrent theme in every survey in this field is *the lack of a proper attitude of management towards maintenance*. What does this sweeping and oft-repeated statement really mean? Findings point to attitudes that range from a lack of interest on the part of management to grave misunderstandings.

In many cases supervisors and maintenance managers complain that "There is little or no interest in what we are doing and how we are doing it." "Difficulties, if they ever want to hear them, are blamed on us and successes are taken for granted." "The only time interest is shown is when a cost-cutting campaign is under way." The overall view is that management has little understanding of the way in which maintenance services are carried out and how the need for maintenance can be anticipated and handled. Neither is there an appreciation of the difficulties encountered in performing a good job promptly, such as identifying the fault, devising ways to repair it, finding what is needed and sometimes overcoming the lack of the right tools, materials and spares.

The two remarks constantly voiced by management are: "You're delaying production," and "You're costing us too much!" The maintenance man can never do right; if he does certain work it costs money, if he does not do it, it will amost certainly cost even more. The misconception of a production-orientated executive arises from two assumptions:

(*a*) production should be kept going at a "tolerable" level of maintenance costs;
(*b*) only turnover of business justifies maintenance expenditure.

If in the light of these two criteria the situation *is* tolerable, management loses interest and they are free to concentrate their efforts elsewhere. Thus maintenance becomes a tolerated activity, "a necessary evil."

It will be useful, therefore, to clarify some points which may lead to a better understanding of this function.

Maintenance is a composite function

No other function in a manufacturing plant, with the possible exception of research and development, involves as wide a range of activities as that of maintenance. Management of the maintenance function abounds with problems of planning, purchasing, personnel, quality control, incentives and technical problems.

Maintenance embraces all these activities as if it were an industry in itself. In some plants, notably in the chemical industry, the importance of this function equals that of production, and its staff often outnumbers production staff. A maintenance supervisor in such plants may not be as hard to find as a good general manager but he needs almost similar qualifications. This, of course, is not the case in small enterprises but the same problems exist there too.

Owing to the wide range of activities within the maintenance function, service cannot be satisfactory unless its activities are as well set up as those of production. Training of workers, planning of work, supply of materials, problems of methods and tooling are all taken care of in the case of production by providing an adequate supporting staff. Is this the case with maintenance? There is often a grave disparity between production and maintenance. Trouble-free production can only be achieved by effective maintenance, and maintenance can only be effective if it is as well organised as production.

The less the demand the better the service

Here we have to explain an apparently paradoxical situation. Whereas in production we aim to maximise the output of products, we try to minimise the need for maintenance services. Assuming a fixed amount of plant and stable conditions of usage, we aim for a constant if not decreasing workload on the maintenance crew. The less our services are called upon, especially for the repair of breakdowns, the more it indicates that the service is effective. While production is forever trying to produce more, maintenance is endeavouring to work less and learn from earlier failures so as to anticipate or avoid such occurrences in the future. Except for preventive activities which have to be pursued relentlessly, maintenance is forever trying to make itself redundant.

This point is often misunderstood and when few breakdowns occur it is credited to other factors. When there are too many, perhaps due to unavoidable circumstances, it is said maintenance is no good and "What else can you expect? Our maintenance people have been sitting on their backsides for the last six months."

The amount of work accomplished is no guide to either the worth of the maintenance staff or to the effectiveness of the service it performs. While regular jobs such as lubrication and inspection *can* be measured in terms of output, the rest of the work cannot. That is where speed of service, ingenuity and initiative are valuable. Quantity of work expressed in figures is therefore no proof for the value of the service.

In view of this argument, management is left without simple guidelines such as those that exist in production to assess how maintenance really performs. Figures seem inconclusive and it is often suspected that there is an attempt to obscure the facts.

The best way to approach the question of assessment is to divide it into three issues, namely: (*a*) how well does the team operate; (*b*) what services are provided for the expenditure incurred; and (*c*) what are the effects on plant and equipment? Control ratios which indicate the results in these areas are discussed later in this book. However, the overall concept of the objective is: "A minimum of good services, required *and* providing at a reasonable cost, for obtaining optimum equipment performance."

Some immeasurable and intangible benefits

Another obstacle to objective assessment is the fact that the benefits of good maintenance are mostly immeasurable and sometimes intangible and the *quality* of the service cannot be assessed except indirectly. Even if we could show management the total value of the service performed as they would have been charged for by a contractor, this would show only the cost of the service but not the many benefits derived.

The effects of good maintenance on the work force, such as improved morale and less accidents, cannot be quantified. Neither can we measure the value of the neat appearance of plant, improved housekeeping and smoother operation in production. Another benefit that is equally elusive but nevertheless important is the improved decision-making process at various levels of management as a result of reliable maintenance data. Since it is impossible to see what is produced we cannot "sell" our services, and management will often be sceptical about the value of the money spent on maintenance.

Although the output of the maintenance department can be quantified in hours, frequencies and cost the total benefits remain immeasurable. For example, what is the value of a routine check which intercepts a serious failure? How are we to know for certain that such a breakdown would, in fact, have occurred? If, as a result of an inspection, we carry out a certain repair, have we just spent money or did we contribute to savings? Therefore, there is no answer to management's quest of a "justifiable expenditure."

"A necessary evil"

Let us now consider the negative attitude that we frequently encounter: "Since we know we have to have it, why not reduce maintenance to the barest minimum?"

As long as no profitability equation exists, short-sighted management will consider maintenance a necessary evil. "Let us consider it as an insurance premium," so the argument goes, "money down the drain, none of it ever to be recovered, etc." The contribution of production planning or quality control is undisputed, while that of maintenance is considered an eternal debit.

A motor-car owner will realise the necessity for replacing a badly worn tyre before a dangerous blow-out occurs. He will go to a well-stocked distributor and get the exact size he wants, expecting it to be there waiting for just this contingency. But the same person acting as a plant manager may object to the stocking of essential spares, intended to reduce the cost and frequency of breakdowns, which are as inevitable here as in the case of his motor car.

Similarly, the resistance of management to adequate maintenance staffing often lies in the difficulty of defining the work content of a maintenance job. Some of the simplest jobs may take hours to complete through no fault of the maintenance staff. The impression persists that "We grossly overpay whatever work is done."

To prove that maintenance is an adjunct to production and not just "a necessary evil" is a matter of semantics, but certainly we cannot conceive of using a car without ever servicing it.

Is maintenance "a bottomless pit" for expenses?

A problem that really haunts maintenance is the difficulty in specifying the *amount* and *intensity* of the service needed for assuring proper plant performance and the proportionate amount of money that should be spent. What is "good enough," and what constitutes over-maintenance? Are there any objective measures for this? These are hard questions to answer. The service we provide can be either above or below the actual needs and there is no indication that this is so at either end of the operation. The expenditure in time, effort and money appears to be undefinable and limitless. Requirements differ widely from plant to plant and from one industry to another. For example, the accuracy of a lathe in a factory for farm implements differs from that of a lathe in the aircraft industry. Evidently, it costs more to maintain machines in top working condition.

Another variable is that of prestige and appearance. The management of a pharmaceutical firm will lay more emphasis on the appearance of their factory and accuracy of their product than the management of a soap factory. Where the safety of people is of prime importance, as in the case of airlines, or the continuity in providing essential services like electricity, no expense will be spared to avoid failures. In another instance, a haulage service may pride itself for being reliable and may go to great lengths to prove its point. But where is the limit? Always more could be done, and when management has only reached part of the way in realising their objectives, they may start to wonder whether the limit of economy has not been passed.

Good systems contain self-regulating devices, which assist in this assessment empirically. When a service is too widely or too closely spaced, events will demand reconsideration of that service. To reach this point a more elaborate system of recording is required in the form of controls which analyse the quality and quantity of results.

Maintenance should be encouraged to try out this approach to reduce costs, instead of embarking on cost-reduction campaigns. Only a good system will indicate whether there has been an improvement.

The "time-lag effect"

Another obstacle in understanding the value of good maintenance is the fact that both the merits and shortcomings of a service are not immediately apparent. The first year of a good lubrication system will pay off in the following years and the effect of poor lubrication is seen when a mishap occurs. In this respect it is hard to give credit or lay the blame for what was done many months ago unless a clear-cut connection does exist, but this rarely happens. Many factors play an interacting role and the adverse ones cannot be identified. This is in contrast to production where a faulty output can easily be traced to either tools, materials or the operators and promptly corrected. An action (such as the replacement of a component or an overhaul) will be seen to have been correct or not only *in retrospect*, and not when it takes place.

Conditions are never stable. Trouble-free running can either be credited to sustained good servicing, to a change in operating conditions or to a new type of lubricant. The non-occurrence of a failure cannot be related to one single cause. More important still, a non-occurrence cannot be recorded. No doubt, figures may show a decrease in the frequency of breakdowns or in their severity, but that could have been due to the latest operator-training programme or to recent improvements in supervision. Because of the time-lag effect maintenance cannot "sell" itself.

There are, therefore, two approaches open to us at this point. We can:

(*a*) relate cause and effect on a time scale; or
(*b*) show a trend one way or another.

It is the second approach that can give us confidence. When proper control is exercised and plotted on a chart a clearer picture will appear.

The foregoing discussion will help the reader to realise why maintenance often operates under adverse conditions. Among them a lack of sympathy shown by management is about the most frustrating.

Difficulties in appraising maintenance can be overcome by instituting a well-organised system, the value of which can be proven by its results. If we do so, we can substitute facts for vague or erroneous impressions and thus pave the way for a better understanding. How to achieve this goal is covered in the following chapters.

Chapter 2

Objectives, benefits and policies

The *objectives* of a maintenance service are dictated by the nature of the enterprise. They vary to a great extent. If, for example, the enterprise is mining, the objective is to minimise the maintenance cost-component per ton of ore. If, on the other hand, the enterprise is a hotel the objective is maximum customer satisfaction. This may entail both a high standard of workmanship and a speedy service. In the case of airlines, reliability is the dominant factor.

Therefore, in order to know how to proceed, the objectives of maintenance must be defined in clear terms. When this is done we can relate the benefits to those aspects that are important for the enterprise. As we have seen, in mining the question of customer satisfaction does not arise. In the hotel business, or in a hospital, the cost of maintenance per one-bed occupancy is the centre of interest.

Perhaps it can all be reduced to the question put by Peter Drucker when he asks: "What *is* our business?"

Policies are a different matter. Since they define *how* we are to achieve our objectives they are not dictated by the nature of the business. They offer a range of possibilities which may reflect the condition of the company and the concepts held by its executives. Thus we can clarify these terms by asking the following questions:

Objectives—What does maintenance have to achieve?
Benefits—What do we expect to gain from good maintenance?
Policies—By what means shall we proceed, within what limits shall we try to work towards our objectives?

Evidently the benefits to be achieved depend on the objectives that are set and the policies indicate how to proceed towards the objectives. These three aspects are strongly interlinked.

In some organisations little attention is paid to these "marginal issues." After all, it could be argued, we can go ahead and carry out repairs and lubrication without defining either the objectives or the means provided for the task. This might be so but it would be like running a taxi service without defining whether we want to satisfy the tourist trade, provide employment, promote the car manufacturers or make a profit.

What kind of service is necessary depends upon a number of factors characterising the enterprise. Table I lists a number of factors which influence the service that is required and the operation of a system that provides the service. Some factors add to the workload while others tend to reduce it. These factors and others must be identified so that a proper system can be instituted.

It would, for instance, be wrong to try to rectify the mistakes of unskilled operators by means of a large maintenance crew instead of embarking on an operator-training programme. Where spares are scarce and difficult to obtain, do we try to make them ourselves—and in that case do we need better facilities—or do we get them by air freight every time?

Table I. Conditions prevailing within the company affect the demand placed upon the maintenance department.

Prevailing conditions \ *Aspects within the maintenance department*	Increased workload	Organisation difficulty	Exacting maintenance	Constant vigilance	Increased costs	Remarks
1. Size of the enterprise	+	+	0	0	+	
2. Degree of mechanisation	+	0	+	+	+	
3. Multi-shift operation	+	+	0	+	+	
4. Trouble-free running required	+	+	+	+	+	
5. Variability of plant usage	+	0	0	0	+	
6. High-quality products	+	0	+	0	+	
7. Poor condition of plant	+	+	0	0	+	
8. Low stock of spares	+	0	0	0	+	
9. Unskilled production operators	+	0	0	+	+	
10. Well-equipped maintenance	–	0	0	0	–	
11. Highly skilled maintenance men	–	–	–	–	–	
12. Regular shut-downs	–	+	–	0	+	
13. Use of contractors	–	0	–	–	+	
14. Lack of stand-by plant	+	+	0	0	+	
15. Corrosive or abrasive materials	+	0	+	+	+	
16. Dusty or humid conditions	+	0	+	0	+	
17. Standardised production plant	–	–	–	–	–	

(+) increase; (–) decrease; (0) no effect.

Considerations such as these will affect our decisions. The mere act of realising the available alternatives will be beneficial. In fact, these are long-range measures to reduce costs. The present list therefore means "If these are our problems, how do we deal with them best?"

The list of factors in the table is by no means exhaustive. Some companies may have other features that have to be taken into account. Some of the factors, such as size and physical arrangement, cannot be easily changed and have to be accepted as they are. Other factors, such as the skills possessed by the maintenance department, can be altered and the aim to do this would be reflected in the objectives set for the maintenance manager and the policies adopted by management.

The contents of this table have to be related to the specific plant under consideration. Evidently, a cement plant will require different services from a garment factory, and the requirements of a commercial laundry are different from a railway company. The factors listed reflect some of these requirements. But whether we think of a fleet of lorries or a battery of punch-presses, the objective is *maximum availability of plant* for "production" activities.

Clear-cut objectives do, of course, exist in the operation of an airline or the maintenance of military equipment. In the former *faultless performance* is essential and in the latter *permanent readiness* is the order of the day.

It must also be borne in mind that this book, as well as the majority of other works on this subject, relates its material mainly to machines and installations. However, most of the procedures also apply to buildings, the grounds and roads. Maintenance of a roof or a fence differs from that of a bearing only in frequency of attention and the craftsmanship required. Stairs and staircases have to be maintained as well as floors and doors. Therefore, whatever is said about mechanical plant also refers to other assets of the company.

ESTABLISHING MAINTENANCE OBJECTIVES

In most companies the objectives of the maintenance function are assumed to be clear and self-evident. Mostly it is in the midst of disagreement that people discover that a wide gap exists between what management and maintenance believe to be objectives of this function. Other views may be held by quality control inspectors or production supervisors. If the disagreement had not occurred, some points would never have arisen. On the other hand the fact that disagreements do not occur does not prove that both management and maintenance are pursuing the same goals. Even where relations seem to be placid on the surface, in certain small ways mistrust may exist, where each side believes the other to be capable of missing some point that the other regards as fundamental.

It is therefore advisable to spell out the objectives of maintenance in clear terms. *Cost objectives* and *operational objectives* are the two main areas that need definition. The realisation that these two groups of objectives differ will explain why management and maintenance are often at loggerheads and why two apparently conflicting attitudes exist. Management is always demanding better results for less money, while maintenance claims that money is too tight. The only way out of this impasse is to clearly define the expected results and the means by which they should be achieved. An assessment of the overall needs of a company in view of its business goals is therefore necessary.

The discussion in this chapter will, therefore, emphasise that management by objectives applies to maintenance just as much as it applies to production. Let us then examine in greater detail the objectives that we have in mind for our particular situation and check our own concept of maintenance objectives in the checklist in Fig. 1. Discussion of these objectives could easily disclose whether they are compatible and workable and, when clearly understood, they would certainly pave the way to a mutual understanding between management and maintenance.

With the passage of time objectives may change. The company's goal may shift imperceptibly to a more demanding product, a faster turnover of volume or a sensitive clientele. How to deal with such issues is, of course, management's task. However, let us make sure that maintenance is not left out of the picture. It must both be kept in-

Which of the following objectives apply in your company?	*Yes*	*No*	*Remarks*
Operational objectives			
1. To maintain equipment (a) in top operating conditions (b) in acceptable condition			
2. To ensure maximum availability for plant and equipment at reasonable cost			
3. To provide service that will avert breakdowns at *all* times and at *any* cost			
4. To extend plant life to the last limit?			
5. To maintain plant and equipment with maximum economy and to replace at predetermined periods			
6. To ensure high-quality performance			
7. To ensure safe and efficient operation at all times			
8. To maximise output over the next five years			
9. To maintain a reasonably good appearance of plant			
10. To maintain a plant spotlessly clean at all times			
Cost objectives			
1. To minimise maintenance expenditure and to maximise profits			
2. To provide maintenance service within the limits of a budgeted amount			
3. To provide funds as a ratio of sales volume production investment			
4. To have maintenance expenditure on the amount of service required by plant and equipment in view of its age and its rate of utilisation			
5. To allow a certain amount of contingencies, tooling and incidentals at the discretion of the maintenance executive			

Fig. 1. A checklist of maintenance objectives.

formed and be allowed to have a say in setting a revised set of objectives. A revision of objectives and policies could take place at two-yearly intervals.

BENEFITS OF GOOD MAINTENANCE

The benefits to be obtained from a good maintenance service fall into the following groups:

(*a*) financial benefits;
(*b*) organisational advantages;

(*c*) technical advantages;
(*d*) human considerations;
(*e*) customer relations.

These groups are displayed in Table II. As mentioned in the introductory paragraphs the assessment of benefits is difficult. Not all the advantages gained can be expressed in figures. The present list shows only practical advantages.

An interesting case of such benefits can be quoted here. A certain telephone company's maintenance force went on strike for a prolonged period. When work was resumed a survey showed that during the strike there had been only the normal number of breakdowns and that operating troubles were no more serious than before the strike. Could we then conclude that the preventive maintenance carried out during the preceding period was unnecessary or that it could be drastically reduced? We would be quite wrong if we did. This event only showed the value of the work performed prior to the strike. The stoppage of service proved how effective the service had been, enabling it to bridge a gap of several weeks without serious defects occurring.

Table II. Benefits of good maintenance.

Group No.	The factors	Expressed in terms of	Units of measurement: Monetary	Units of measurement: Non-monetary	Important for
(a)	*Financial benefits*				
	1. Extended plant life	Book value	£	–	Utilisation of capital
	2. Uninterrupted production	Higher plant availability	£	hrs	Utilisation of capacity
	3. Improved quality of production	Reduced scrap and inferior grades	–	%	Value of output
	4. Reduced production delays	On-time deliveries, less delay penalties	£	freq.	Customer relations
	5. Reduced costs of repairs	Maintenance costs	£		Maintenance economy
	6. Less stand-by plant and spares	Inventory carrying costs	£	–	Capital utilisation
	7. Improved equipment replacement	Lower plant cost per unit of product	£	–	Unit costing
(b)	*Organisational advantages*				
	1. Co-ordination between production and maintenance	Improved understanding			Internal planning
	2. Manpower planning	Unproductive time	£	hrs	Manpower utilisation
	3. Planning of deliveries	Cost of delays	–	–	Plant utilisation
(c)	*Technical advantages*				
	1. Improved equipment suitability	Better choice of machines			Optimum production faciliti
	2. Build up of technical data	Better and more accurate information			Standard of technical service
	3. Improved maintenance schedules	Minimum maintenance costs			Plant availability
	4. Improved plant condition	Performance and reliability			Operating efficiency
(d)	*Human considerations*				
	1. Increased safety	Losses due to claims and less production costs	£	freq. & severity	Production economy
	2. Improved housekeeping	Tidiness of shop floors			Workers' morale
	3. Less friction, better relations	Harmonious relationships			Staff relations
(e)	*Customer relations*				
	1. Reliable delivery dates	Improved reputation	–	–	Sales promotions
	2. "Showcase" housekeeping	Better public image	–	–	Company image

A few of the good effects of maintenance become noticeable shortly after the introduction of new procedures. At first the internal procedures of the maintenance crew are affected. As soon as schedules are explicit and well laid out, work allocation within the maintenance department becomes easier. Workers know their instructions and less time is wasted on enquiries and clarifications, and waiting time between jobs is sharply reduced.

Next, the presence of maintenance personnel in production areas becomes more frequent and more purposeful. Production supervisors and foremen will accept this increased presence with mixed feelings. Some may consider it as a passing wave of interference with shop-floor work and some may welcome it as a long-overdue effort, expressing the hope that it will not be short-lived. It usually takes some time for production personnel to realise that they are getting more and better service and that ultimately the frequency of breakdowns will decrease. The response to calls will be more prompt now that maintenance is operating efficiently.

The accountants, however, will be having a hard time. If they have been alert they will have noticed increasing costs during the previous few months as more and more items have been ordered and purchased for the maintenance department. The workshops that were sadly neglected in the past have had to be fixed up, desks have had to be bought for the craft foreman and new forms ordered from the printers. As more work is carried out by the maintenance crew, consumption of materials and spares rises at what may seem to the accountants an alarming rate. Figures will show that all costs have risen except the maintenance payroll. The question may arise, "Why do these people suddenly spend so much?"

If management had been warned about this at the time they gave their original approval, they will understand. If not, they may need the reassurance that this is only temporary and part of the project. Sometimes management may be impatiently awaiting positive figures that show conclusively the benefits achieved—this is a great obstacle to understanding.

The overall benefits are the comprehensive result of all points outlined in Table II. Chapter 10 of this book discusses controls that indicate whether improvements in some areas have been attained. However, some results do not necessarily indicate benefits. For example, more jobs are being finished than before and more machines are being serviced, or is it really that more defects are being discovered? Is there benefit in this? Were all these jobs really necessary? Only time will tell.

Financial benefits can only become evident if data are available and controls are exercised to interpret and evaluate them. Even so, long-range benefits cannot be produced overnight although some short-term benefits may soon become evident. Where production is measured by daily output and when the existing frequency of mechanical trouble is a limiting factor, output figures will soon indicate improvements and then benefits are easily measurable. Losses due to down-time, however, are not so easy to calculate and recurrent repairs also affect the life of the equipment. This makes the total balance of benefits difficult to calculate.

The technical advantages are often very remarkable, but they are appreciated only by the technical staff, and management may not be aware of their importance. In this respect, the benefits reaped will prove to be in direct proportion to the effort invested in the system. Extracting information and interpreting facts is sometimes neglected in the rush of activities. Therefore the right steps should be taken to ensure that this area is fully exploited.

In the field of human relations it may appear that the personnel department has a monopoly. For instance, who would allow maintenance a grudging credit for improving safety? Of course, safety posters and markers are probably put up by maintenance as a result of an initiative by the personnel department or the safety committee, but little consideration has so far been given to the fact that maintenance does contribute to improving the working environment of production workers.

Other criteria for judging the usefulness of maintenance work may be plant appearance. Evidently there is no *tangible* benefit here, and there are always those who ask, "What is the point of it?"

The benefits that really matter take a long time to become evident and even longer to sink into people's consciousness. The fact that emergencies become less frequent and that less of management's time is taken up in chasing spares and in seeking solutions, is a submerged benefit that cannot even be put on paper, however important it is for the smooth conduct of an enterprise. When controls are available, figures that show an increase or a decrease of certain variables become important. Information will be produced to enable management to make decisions on matters of personnel, training, tooling, spares supply, stocks and, above all, replacement of obsolete equipment.

It must be clear by now that sporadic efforts to improve the maintenance service will not achieve the same results. Only a system, built up and integrated into the company can attain lasting and far-reaching results. Such a system must contain self-regulating devices and controls that indicate progress. Subsequent chapters will describe how such a system can be set up.

MAINTENANCE POLICIES

Policies are guidelines which assist in achieving the defined objective of a function. They indicate a course of action to be followed to cope with situations as they arise. To check the validity of a policy we can ask: "Does it help us to deal effectively with an existing problem?" or "Does it deal smoothly with a recurring situation?"

A study shows that maintenance policies are rarely defined and very rarely exist in written form. To quote the American Management Association, *Control of Maintenance Costs*, Research Report No. 41, New York, May 1964:

> "Maintenance policies have usually evolved from experience in a rather haphazard fashion rather than being designed to implement specific maintenance objectives. Relatively few of the companies . . . have written maintenance policies. Existing largely in the minds of maintenance supervisors, uniformity and continuity in policies tend to be lacking. In part, this situation is attributable to lack of management attention to maintenance. A more important explanation may be found in the difficulties encountered in applying quantitative analysis to establish maintenance objectives."

Policies which relate to the operation of maintenance can be grouped as follows:

(*a*) scope and limits of maintenance;
(*b*) type and level of service expected;
(*c*) responsibilities to management;
(*d*) personnel practices;
(*e*) trade functions and trade-union contacts;
(*f*) budget and financial controls.

These policies can be presented in chart form as in Fig. 2 which shows the interrelationships that exist between the various functions of the company. Each of the numbered relationships must have detailed policies so that every eventuality can be dealt with expeditiously.

The advantage of defining and using policies is that management can convey its intentions to all ranks. Policies both expand *and* limit authority since they ensure that

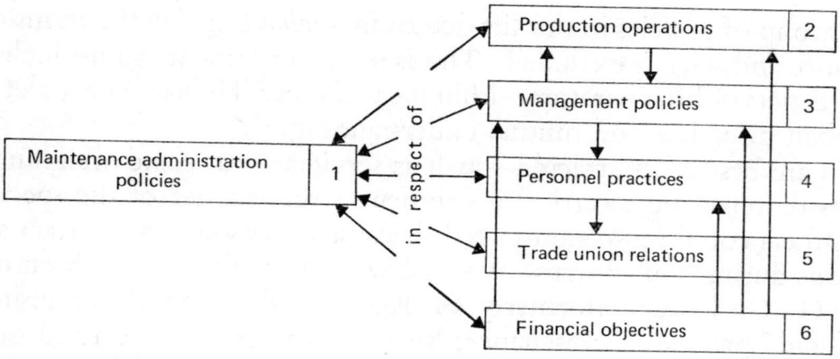

Fig. 2. The interrelationship between maintenance policies and other groups of policies within the company.

people act within the whole range allowed by the definition. In the absence of policies people will tend to emphasise their limitations, so that they do not risk overburdening themselves nor extend their responsibilities too far.

The functions that are expected to be performed by maintenance are not self-evident. What often happens is that some functions are explicitly only stated in part, some are assumed, some are assigned to maintenance in the absence of other takers, and some seem to appear out of nowhere. There is nothing wrong with jobs being done that need to be done, but it is important to realise that when too much of an extraneous load is carried by maintenance it will distort the picture. When this happens neither management nor maintenance can see the true facts. It is only by means of policies that we can define what maintenance is actually expected to perform in a given enterprise.

The following example may help to illustrate this point. In a certain railway repair shop, employing about 500 direct workers and using a great deal of mechanical equipment, the maintenance section numbered twenty-eight people, namely 5·5 per cent. Although this figure was considered rather low in view of the condition of the plant it was obvious that service could have been made more effective by better planning.

Since service was evidently lagging far behind the needs, management ordered an investigation to be made. It was found that the group included four crane-operators who also serviced their cranes, one ambulance-driver, two tool-store attendants, one refractory-bricklayer, one sling-splicer, one band-saw sharpener and a boiler inspector. When the working time of the remaining seventeen workers was analysed, such a vast range of auxiliary jobs was discovered that only a monthly average of 1,600 hours of mechanical repair jobs could be traced. This is equivalent to the total working hours of eight or nine workers. The time on extraneous jobs was spent on repairs in the works canteen, on office equipment and on production jobs. The true picture differed greatly from the original impression. If policies had been laid down maintenance manpower could have been utilised much more effectively.

What should be the scope of activities?

The first group of policies in our list determines *what is within* the terms of reference of maintenance and what is excluded. This is not an argument against including floor-cleaners, gardeners or lift-operators within the team, but if it has to be so, let it be on the basis of explicit directives and mutual understanding.

Some jobs are best subcontracted—such as servicing scales and weighing machines, tarring roofs or repairing electrical instruments—on account of the specialised skills required. Others could be subcontracted, because of seasonal loads, such as overhauls of installations during shut-downs. A worthwhile contribution has been made in this respect by G. G. Corder (formerly of Kodak Ltd.) who distinguishes between "domestic" and "process" maintenance; such groupings can be carried out in several ways. The most easily identifiable functions to be performed by the maintenance department are discussed under "What does maintenance really do?" in Chapter 3.

The second group of policies refers to *the amount and intensity* of the service that maintenance is expected to provide to answer the need of production. These policies would reflect the objectives of maintenance. Just short of being a financial commitment on the part of management, these policies indicate how much service is expected to be "delivered." The *level* of service, its intensity and thoroughness depends on the balance between the cost incurred and the functional objectives of the company. If cost-saving objectives predominate, the level of service may be restricted. There are, however, two aspects of the operational level: (*a*) the performance of equipment and (*b*) its appearance. The former lends itself to quantification, *i.e.* the mechanical condition which manifests itself in the quantity and quality of production output and in the degree of trouble-free operation. The appearance aspect is determined mainly by company policy. Companies aiming at improved customer relations may emphasise a "show-case" level of appearance. It is often expected that by doing so employee attitudes may also be favourably affected.

Thus standards for the level of maintenance for achieving both high performance and good appearance have to be clearly laid down by means of policies. These will allow a better interpretation of maintenance results and costs. In fact, the cost of keeping a high level of appearance should be provided either within the normal maintenance budget or separately in an account which can be called a "good housekeeping" account. Since the effects derived would be directed towards improving customer or community relations it could perhaps be taken, in part, from an advertising budget. The mere exercise involved in defining policies and adopting them will have beneficial effects both on management and on maintenance. A better co-operation will develop and many areas of misunderstanding will be clarified.

The largest number of conflicts between maintenance and other departments arises in the field of priorities. There is as yet no statistical proof for this assertion, but any consensus of opinion among people in this field will confirm it. It is a fact that inefficient maintenance is carried out mostly in a rush, especially when breakdowns occur and there is pressure for immediate attention. When this happens a crisis can easily occur; people tend to regard their own troubles as more urgent than anybody else's and the clash of opinions can become quite severe.

The best way to avoid such friction is by establishing priorities. We are often confronted with situations where decisions rely on agreed compromises. This is especially so when alternative solutions, more or less expensive, are also available. The question of priority does not relate only to urgency as the following example will show. An old medium-sized compressor keeps breaking down. The lack of compressed air is a nuisance in production, but the twenty people affected manage to get by without it for short periods. On the other hand, the extractor unit in the spray booth affects two workers only. But because it is not easily accessible it may take a day or two to clean the vents and replace its failing motor. To eliminate repeated trouble both cases need radical solutions which will divert time from regular maintenance. Should money be spent first on an ageing compressor or should time be spent on an extractor which is due to be replaced? Considering all the possible solutions which of these should be given priority?

It can be seen that the priorities to be established relate to:

(*a*) urgency of attention when simultaneous emergencies arise;
(*b*) amount of work to be spent on certain jobs and equipment;
(*c*) replacement schedules and guidelines; and
(*d*) priority of routine service *versus* irregular jobs.

The next group of policies relate to the *dependence of maintenance* on management. Questions often arise about the authority to act, to approve a certain job and to order replacement parts. It may be trite to assert that maintenance managers and supervisors must be motivated, and in any case must not be de-motivated. By restricting authority, management denies the maintenance executive the opportunity to prove his competence to his own staff and, to the company, his ability to manage. There are other implications as well. Establishing proper channels for authorisation breaks the sequence and causes delays, thereby giving rise to job frustration and resentment against management. Harm resulting from this can easily outweigh the benefit of any savings achieved.

Nothing is as frustrating to the maintenance supervisor as trying to do his work burdened with poor equipment and lack of spares and still to be blamed for excessive costs. This situation arises when budgets are set arbitrarily or not at all, or when expenditure requires a prolonged approval procedure.

Policies must be set that enable the department to operate within the required *limits of expenditure*. Provision must be made to take into account the true condition of the plant and its usage, the purchase of spares and materials and tools for maintenance.

Policies relating to the budget do not determine the amounts to be allocated, which will vary, but the limits in the form of operational ratios or other bench-marks. It should also be decided in what way control is to be exercised by management. This is fully discussed in Chapter 10.

Policies must be set that enable the department to operate within the required *limits of expenditure*. Provision must be made to take into account the true condition of the expressions of authority, such as using one's best judgment or one's initiative when unforeseen contingencies occur.

Another point to be covered in this group of policies is the *nature of reporting* practices. This will achieve the double effect of requiring maintenance to report and at the same time obliging management to take an interest in maintenance. A by-product of reporting is the exercise of controls. Policies would provide answers to such questions as: "What activities should those controls reflect?" or "What is the accuracy demanded and the frequency of reporting?" We may notice that *if* this approach is followed most of the problems of management's attitude to maintenance may be solved.

Policies regarding *maintenance personnel* must deal with the following points:

(*a*) strength, as related to the scale of operations;
(*b*) selection, training and promotion procedures;
(*c*) trade skills of craftsmen;
(*d*) supervisory strength and practices;
(*e*) pay scales and incentives;
(*f*) safety programmes.

As we have seen the adoption of adequate policies provides an operating charter that will serve management, production and maintenance. The establishment of policies requires adequate information regarding the plant and the necessity for co-ordinating existing needs with available means. Working out the details and incorporating them into an "operating manual" will mark a milestone in the work of the maintenance team.

Chapter 3

Organisation of the department

The establishment of an efficient maintenance service requires the combined efforts of all concerned, every member of the organisation must be receptive to it. It is up to management to issue the relevant directives and to give their full support to the maintenance executive. The effectiveness of the system depends on sound interaction of the main participating departments namely: plant engineering, production, cost accounting, stores, purchasing and personnel. Since each of these has a particular point of view, management may have to act as the arbitrator and to incorporate details in the plan which will lead to the best overall results.

Smooth operation of the service requires the following aspects to be clearly defined and understood:

(*a*) the position of maintenance within the company;
(*b*) the internal organisation of the department;
(*c*) the functions and responsibilities of key personnel.

With these definitions established a good foundation has been laid for harmonious and efficient inter-departmental collaboration. Obviously the larger the enterprise the more important it is to define duties, limits of authority and responsibilities.

Friction and misunderstandings will arise unless a satisfactory structure is established by the combined efforts of all concerned. We all know of cases where arguments arise repeatedly over the same issue. It is then we wish that fixed guidelines had been laid down that could not be disputed and would resolve certain problems once and for all. What we really need is a "charter of operations" that clearly lays down the terms of reference of every function and position. It is useful to remember that when people are directly involved in making decisions they will take more interest in the results of their decision-making. This is true of both management and of other departments.

We know from experience that it is useful for the maintenance manager to hear his problems discussed from different points of view. It is often claimed that management and other departments do not understand the maintenance point of view, but only rarely is anything done to overcome this difficulty. Finding the time to do it seems the most frequently cited excuse, but any effort in this respect will be richly rewarded.

MAINTENANCE WITHIN THE OVERALL STRUCTURE

The position of maintenance within the overall structure of the company has a great impact on the effectiveness of this function.

Depending upon its position in the company structure, maintenance may be able to attract more co-operation from all functions, obtain more technical assistance or create more interest on the part of management. If, for example, it is subordinate to production it can only appear as one unit among many. When this is the case maintenance requirements become secondary to those of production.

The basic organisational guidelines for the maintenance function are as follows:

(*a*) All activities relating to installation of plant and maintenance of equipment should come under one authority. This should be the case even in small companies where either the chief engineer will be in charge of maintenance or the manager himself.

(*b*) The chief maintenance executive should report to as high an authority as is feasible, preferably the plant manager. He is the person to ensure a balance in the everlasting clash of interests between production and maintenance.

(*c*) An organisation chart is helpful in establishing the position of maintenance and other functions. In some cases certain persons may try to dominate or block relationships. A chart will help to overcome this.

(*d*) A chart should not be regarded as Holy Writ since it presents only the formal side of the system. Informally, relationships will depend on personalities. Discussions leading to the adoption of a chart are always helpful. Above all it is important to get management's approval of the final chart and the backing for its implementation.

(*e*) The standing of all the maintenance staff is greatly enhanced by the status of their chief. The higher the authority the maintenance manager commands, the more favourable the effect on his team.

(*f*) Maintenance should *not* be responsible to production. It is a primary function which cannot be managed as a part-time activity of the production manager. If in certain departments repairmen are required on a full-time basis they can be so assigned and still be responsible to maintenance. This is one aspect of decentralisation.

(*g*) There is no universal organisation chart which can fit all situations. Every company has features peculiar to itself and these, too, change sometimes. Figure 3 assumes that, as according to Fayol, there are five basic functions in the enterprise subordinate to management. Practices differ widely and some functions may not be separately identified in small plants but Fig. 3 can still serve adequately as a chart of reference.

(*h*) It is accepted nowadays that plant engineering combines the function of planning for improvement and the expansion of installations with their upkeep. These functions complement each other. The supervision of power plant and utilities, *e.g.*

steam, water, compressed air, are normally assigned to maintenance, sometimes under a separate foreman. Maintenance stores come undisputedly under plant engineering.

(*i*) Purchasing of spares, as well as maintenance of transportation vehicles, must be assigned according to the size of the enterprise and a designated degree of expediency.

(*j*) Figure 3 represents the basic relationships that *could* exist between the various technical functions and should be developed as changes occur in the situation.

It must again be emphasised that drawing up an organisation chart of this sort should involve all departments. In some cases, because of the small size of the enterprise, what

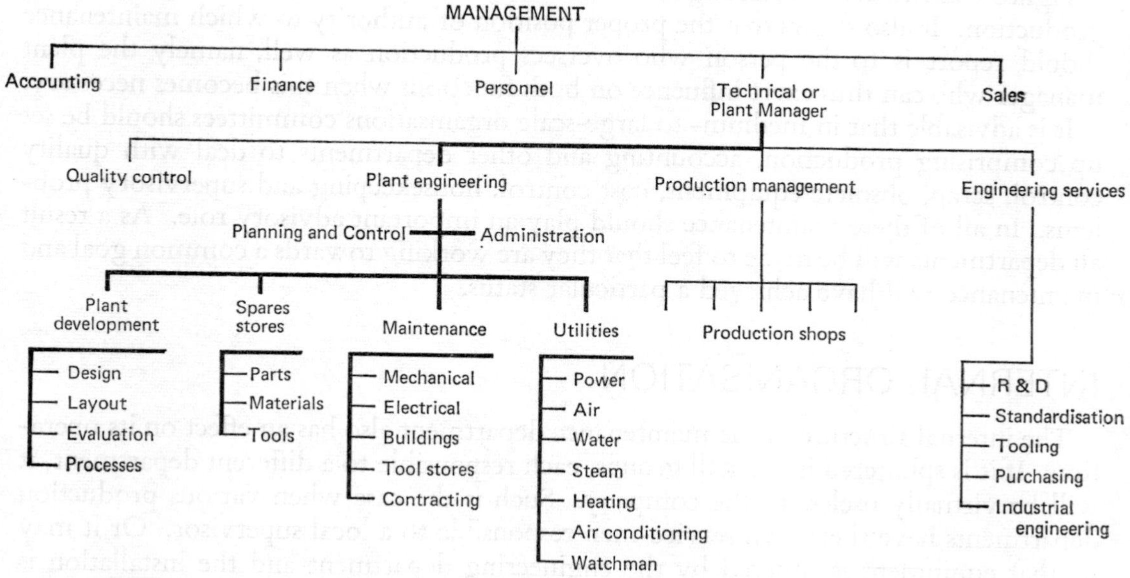

Fig. 3. The position of plant engineering and related departments within the organisational structure.

some people would refer to as "all this nonsense about charts" is not justified. It is then advisable to discuss these matters until an agreed framework of procedure is established and to refer to such a framework when the need arises.

Changes in the situation arise when new processes are introduced or when the enterprise expands. It may then become necessary to revise the organisational structure.

If we accept the structure of the chart in Fig. 3 a few more persistent problems can be solved. These are as follows:

(*a*) Maintenance stores become the direct responsibility of the plant engineer and if stocks occasionally run out, he has nobody to blame but himself. Naturally it implies that supplies are ordered and purchased according to the agreed budget. This will eliminate the passing on of responsibility to anybody else.

(*b*) Close co-ordination is possible between the purchase and installation of new plant and its subsequent maintenance. The plant engineer in many firms also develops special tooling and handling equipment. On a smaller scale, however, no distinction at all is made between these functions.

(*c*) From Fig. 3 it becomes clear that everything that has to be maintained comes within the plant engineer's jurisdiction in one way or another. Some of the disputed areas are power installations, transport and handling equipment, production tooling, building maintenance and instrumentation. Solutions for providing services for these groups may be found in subcontracting or in subdividing the maintenance department into specialised teams.

Figure 3 shows that accounting serves maintenance in much the same way as it serves production. It also shows that the proper position of authority to which maintenance should report is to the person who oversees production as well, namely the plant manager who can thus exert influence on both functions when this becomes necessary.

It is advisable that in medium- to large-scale organisations committees should be set up comprising production, accounting and other departments to deal with quality control, scrap, obsolete equipment, cost control, housekeeping and supervisory problems. In all of these maintenance should play an important advisory role. As a result all departments will be made to feel that they are working towards a common goal and maintenance will have achieved a particular status.

INTERNAL ORGANISATION

The internal structure of the maintenance department also has an effect on its operation. If it is splintered into small groups, each responsible to a different department, it will be virtually useless to the company. Such is the case when various production departments have their own service-men responsible to a local supervisor. Or it may be that equipment is ordered by the engineering department and the installation is carried out by outside contractors, by-passing the maintenance department altogether. Such cases still exist but they are becoming few in number.

There used to be a time when the recognised division within maintenance was between crafts. In smaller companies even today all "repair" crafts came under one foreman. It is sometimes proposed that organisational grouping can be either by areas or by crafts or a combination of both. Today, however, with the expansion in the size of factories and the consequent introduction of many complicating factors the department can be safely split into more specialised groups.

In order to arrive at the best team organisation the functions of maintenance can be classified as follows:

(*a*) trades, *e.g.* mechanical, electrical, building, instrumentation, etc.;
(*b*) types of service, *e.g.* lubrication, inspection, repairs, overhauls, etc.;
(*c*) areas or groups of equipment;
(*d*) planning aspect of services, *e.g.* emergency repairs, routine or regular service, fixed location assignments (boiler-room), etc.

The organisational structure in any one factory will usually represent a mixture of these since it is rarely possible to prescribe specific solutions. The examples in Figs. 4, 5 and 6 illustrate some of the more common cases.

As these examples show, the variety is unlimited. Specialised crews must be fitted into the appropriate set-up. The repair of instrumentation, safety devices and automation controls requires special solutions depending on the amount of work they create. In representing the various functions on a chart we may have to contravene some charting conventions. There is a rule, for instance, which says that all functions on one level should be homogeneous, *e.g.* trades or areas only, and to put power-house, packing department, maintenance, mechanical repair and lubrication on the one level

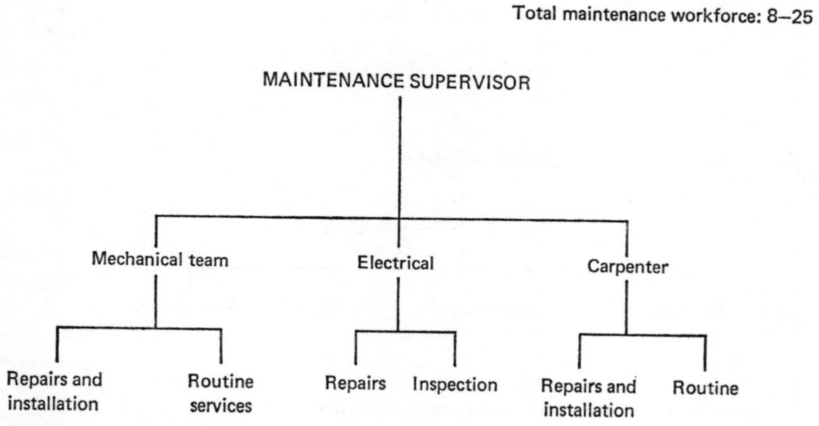

Fig. 4. Maintenance-team organisation in a small firm.

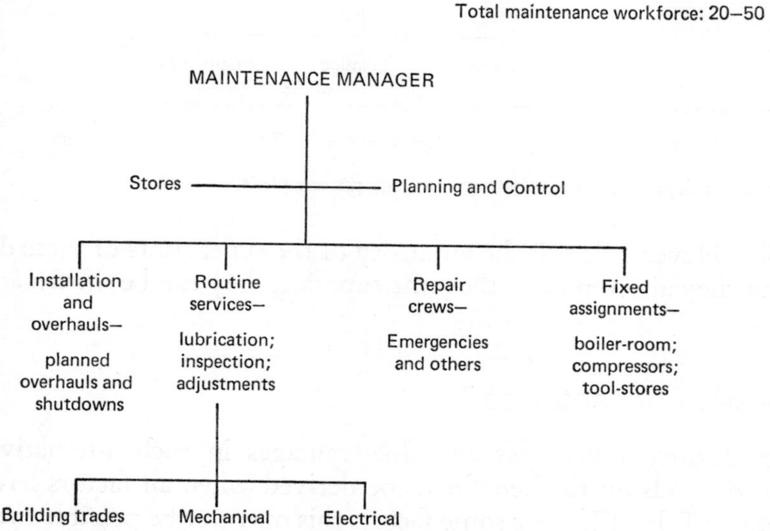

Fig. 5. Maintenance-section organisation in a medium-sized manufacturing company.

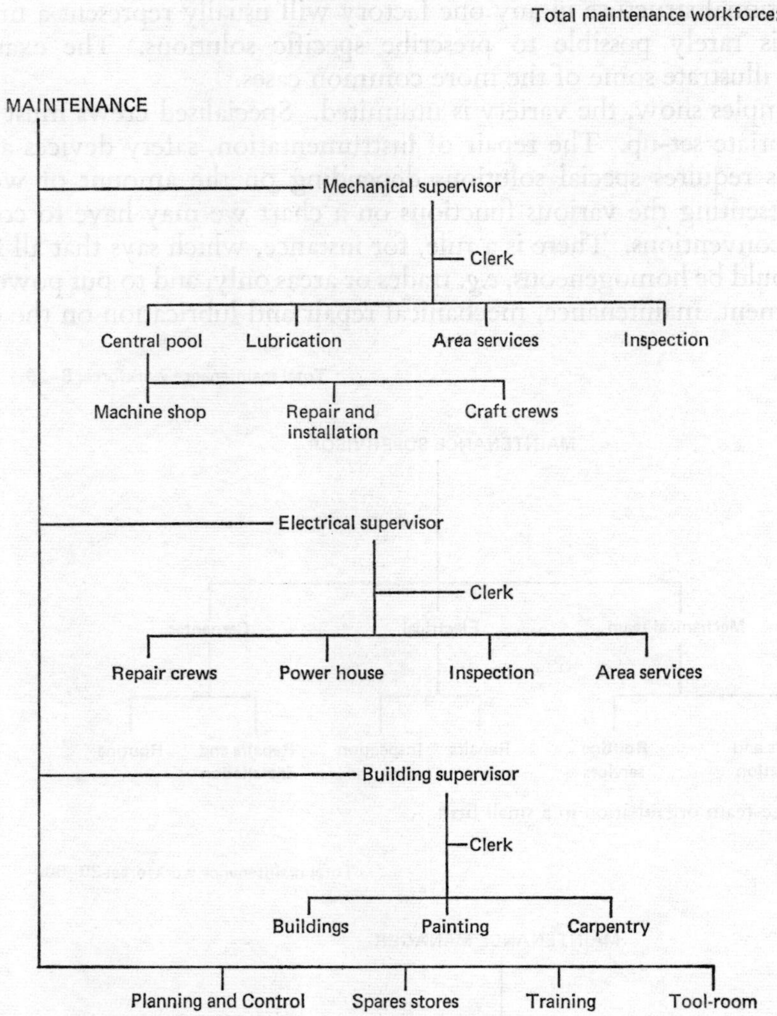

Fig. 6. Maintenance-department organisation in a large process plant.

is undesirable. Nevertheless, if the authority of the supervisors of these departments is the same, and they also report to the same superior, they can be shown accordingly on the chart.

To centralise or decentralise?

There are distinct advantages and disadvantages in each alternative. The final arrangement depends on the benefits to be derived when all factors have been taken into account (*see* Table III). For some factors this may not be possible even though we are aware of their importance.

The overall merits of decentralisation are decisive where both speed of service and specialised know-how are required. Where expensive materials are processed or machine time is very valuable immediate service is essential. When repairmen are constantly away from the central shops while they are in fact required in other locations, the situation should be investigated. An analysis of calls will indicate whether decentralisation would be beneficial. In any case, a decision should be based on an objective evaluation of the known facts, such as the specialised requirements of certain hops, the walking time to location and the frequency of calls.

Table III. Advantages and disadvantages of decentralisation.

	Advantages	*Disadvantages*
Centralised	Ease of planning Ease of supervision Well-equipped shops Effective control of manpower	Longer walking distances No specialisation possible
Decentralised	Speedy service Specialised know-how Constant attention to plant Less paperwork	Duplication of tools Dual authority Poor records Poor skill utilisation

Decentralisation is a solution which should be selectively applied. This may be contrary to principles of organisation as quoted in most textbooks which prescribe a homogeneous structure. Within the maintenance department, however, it will harm the organisation in no way if only one area or function is decentralised, as evidenced by a traditional boiler-attendant who operates on his own away from the group.

A problem which often arises is that the tradesmen assigned to certain areas may have two bosses. This conflict is resolved by limiting the authority of the local supervisor. He would prescribe priorities for repair jobs since they affect production, but would not otherwise interfere in the performance of maintenance. In most cases a better co-operation will exist between production and maintenance owing to closer contact and a better understanding of each other's tasks.

RESPONSIBILITIES OF THE KEY PERSONNEL

Defining the responsibilities of key personnel is essential for ensuring their effective work. Let us look at examples, in Figs. 7, 8 and 9, of job descriptions for the works or plant engineer, the maintenance foreman and the maintenance supervisor.

These examples are not exhaustive and may be adapted to any given situation. A periodical revision should be carried out to prevent job descriptions from becoming stale or out of step.

A function which is often missing in smaller teams is that of a clerk-planner. In a maintenance department of twelve craftsmen such a job may prove well justified. It must be borne in mind that most *production* workers have their regular assignments,

1. *Position and title*	Plant engineer (works engineer)
2. *Department*	Plant engineering and maintenance
3. *Responsible to*	Plant manager (works manager)
4. *Basic function*	Administration of plant engineering
5. *Immediate subordinates*	Maintenance supervisors, plant utilities manager and plant development
6. *Duties*	To manage the installation of new equipment and provide for suitable maintenance services
	To develop effective maintenance methods and procedures and to keep practices up to date
	To plan and submit yearly budget requirements
	To provide management with control data
	To maintain working relationships with all departments and other service functions
7. *Responsibilities*	To supervise implementation of the maintenance programme
	To achieve the aims of his functions within the allocated budget
	To provide at all times safe working conditions for his subordinates
	To plan the installation of new equipment with provision for convenient access for servicing
	To utilise all resources in the most economical way
8. *Authority*	To act within his engineering capacity
	To instruct his staff in their respective jobs
	To set down the scope and limits of his subordinates' jobs
	To organise the plant engineering and maintenance activities
	To act within the accepted expenditure practices
	To report to management on the activities of his department
	To advise on the purchase and replacement of equipment
	To act within his function so as to achieve maximum efficiency

Fig. 7. Job description of the plant engineer.

work orders, inspectors, workteams, tools and materials, but in maintenance these elements are never constant; they have to be combined afresh for every job. The craftsmen's time is as expensive as breakdowns are costly, and planning and scheduling can be considered, therefore, as a profit-making activity.

The issuing of job cards, the recording of work done and the time spent on jobs, the issue of service schedules and the follow-up on the availability of spares are routines that should exist in even the smallest operations, and to expect a foreman to deal with them all would be both unrealistic and detrimental as his normal duties would inevitably suffer.

Job descriptions such as these enable a person to act with confidence within the defined limits. They should be well prepared; care should be taken to ensure that there are no gaps in procedures and that there is no overlapping in duties between related functions; and they should also undergo periodic revision.

1. *Position and title*	Maintenance foreman or craft foreman
2. *Department*	Maintenance section
3. *Responsible to*	Maintenance supervisor or manager
4. *Immediate subordinates*	Craft workers and apprentices
5. *Basic function*	Supervise repair and service teams
6. *Duties*	To assign jobs to tradesmen, to follow up on progress and to inspect finished jobs
	To assist and train workers in the performance of their work
	To plan each day's work for all workers and to plan ahead
	To balance teams according to workload
	To determine priorities
	To maintain in efficient condition workshops, tools and stores
	To carry out the prescribed recording procedures
7. *Responsibilities*	To submit reports on the use of workers' time, on attendance and overtime
	To use tools and materials efficiently
	To ensure safe working conditions
	To carry out management's instructions and policies
	To follow standard practices and procedures
8. *Authority*	To withdraw necessary materials from stores
	To accept or reject work
	To advise on training needs and promotions
	To deal with grievances
	To approve overtime

Fig. 8. Job description of the maintenance foreman.

With an increased scale of maintenance operations, it is advisable to have similar job descriptions for other key persons in the department. Since there is no guarantee against people moving on to other jobs, or retiring, job descriptions are a great help when new persons take over. They also provide a clear-cut framework for people seeking promotion. The title coupled with wider responsibilities often serves as an incentive for ambitious workers. Eventually it will be found that job evaluation and a sound salary structure become a necessity. It is wise, therefore, to anticipate such situations by laying the foundations at an early stage.

Strength of the maintenance workforce

Strength of the workforce depends on factors such as the work the team is expected to do, their degree of training and craft skills, the amount and condition of equipment, their motivation and the quality of supervision. No generalisation is possible.

1. *Position and title*	Maintenance supervisor	
2. *Department*	Maintenance section	
3. *Responsible to*	Maintenance manager	
4. *Immediate subordinates*	Mechanical, electrical and building craft foremen	
5. *Basic function*	Management of maintenance services	
6. *Duties*	*(a)*	To supervise installation, maintenance and overhaul of all mechanical and electrical equipment on the premises
	(b)	To organise the maintenance procedures
	(c)	To co-ordinate his section's work with production
	(d)	To utilise information regarding all plant to the best advantage of the company
	(e)	To report to management regularly
	(f)	To assist production in the development of special tools
	(g)	To maintain proper discipline in the shop
	(h)	To supervise craft foremen and training of craftsmen
7. *Responsibilities*	*(a)*	To maximise availability of equipment for production
	(b)	To ensure safe working conditions, and the provision of safety equipment
	(c)	To ensure an adequate supply of tools and materials
8. *Authority*	*(a)*	To direct the work of his subordinates
	(b)	To authorise repairs and overhauls involving costs of up to £200
	(c)	To sub-contract jobs up to £50
	(d)	To order spares, tools and materials as required with the allowed budget
	(e)	To advise on the replacement of equipment
	(f)	To advise on the yearly budget allocation
	(g)	To authorise overtime work, training and promotion within his section
	(h)	To advise on the employment of maintenance personnel

Fig. 9. Job description of the maintenance supervisor.

Many attempts have been made to establish bench-mark figures from averages in various branches of industry but comparisons hold true only to a limited degree. One enterprise can be compared with another given similar conditions, namely the type and amount of equipment, production manpower, the products, the quality level expected and so on. In these circumstances it is possible to compare three maintenance workers per hundred production employees in one plant with four or five in another.

Although recently a number of thorough surveys have been carried out,[16–18] they show that even within the same type of industry large fluctuations still exist. This may be due to the way in which maintenance service had been established and accepted. An exceptionally large crew size or a very small one may reflect not only the age and condition of the equipment but also management's desire to have them differently classified. It would be very hard to relate such variables to the effectiveness of the crew or to its workload.

The best guide for justifying the size of the maintenance team results from answers to questions such as: "What is the average workload?"; "Are all hands well occupied?" and "What is the degree and amount of attention the equipment is getting?" Unless there is an attempt to assess the amount of work in a particular enterprise, external comparative figures are not particularly relevant.

In broad terms, it is the degree of mechanisation, the complexity, age and condition of machinery that will influence the size of the team. In this respect, the classification of industries into heavy process, light process, heavy fabrication, light fabrication and benchwork (*SIC—Standard Industrial Classification Manual*, U.S.A.) can serve as a good guide. Subsequent chapters dealing with planning of the workload, supervision and controls will provide more specific answers and aspects of indirect maintenance staff are discussed under "Supervision" in Chapter 9.

Until we actually carry out some analytical spadework, any conclusions that we may draw will be purely conjectural. One rule of thumb can be applied: how many people did we have two years ago, one year ago, and how much has our company developed in the meantime?

WHAT DOES MAINTENANCE REALLY DO?

What *constitutes* maintenance work?
What are the functions of *our* maintenance department?
What do *we* expect of maintenance?
Each question has its own answer and they should be kept separate. However, people often treat them as if they were one and, therefore, interchangeable. This is not the case and we shall provide the correct answers one by one.

> *"Maintenance" refers to all the activities which assist in keeping plant and equipment in good condition.**

Apart from the obvious jobs of repair, lubrication, overhauls, inspection, and replacement of parts, there are others which are less immediately evident. Among these duties we find:

Training of the maintenance staff, including the foreman.

* The German DIN Maintenance Committee has proposed the following alternative: "Maintenance—measures to maintain the required condition of buildings, installations and technical equipment", quoted by Maintenance Management Newsletter, August 1, 1973, published by the Maintenance Advisory Service, Silver Glade, Gasden Lane, Witley, Surrey, GU8 5QB. Other definitions from Australia, Japan and Italy are also available from the same source.

Training of production operators.
Testing parts for suitability.
Planning of servicing schedules.
Improvements and modification of plant.
Protective painting.
Production of spare parts.

These activities share one objective, namely, *maintaining the physical assets in good shape and appearance*. It is important to remember that preventing adverse conditions from arising is as important as patching up. This definition *excludes* service by cleaning security staff but *not* maintenance of grounds and gardens.

The installation, moving or dismantling of equipment are not strictly within the meaning of maintenance; neither is the manufacture of replacement parts nor the modification of machines to improve production. More accurately expressed, they are *plant engineering functions*. Since these functions merge in small plants they have come to be regarded as one until a sharp growth in size demands that they should be separated. Whether at that stage we include in the maintenance department the people concerned with plant development, is an organisational problem to be solved in a manner that suits the particular enterprise.

Does this distinction seem to split hairs? It does not, in fact, when one realises that plant *development* may sometimes be done by a limited staff at the expense of regular maintenance. Some maintenance teams are overloaded at times with construction and installation jobs which may last for weeks or months. The number of people carrying out true maintenance is thereby reduced and the service suffers accordingly. If at this period we were to relate man-hours of direct production (or units of production) to man-hours of true maintenance, we would get a low figure, *e.g.* five hours of maintenance to two hundred of production. When construction jobs decrease, the hours spent on maintenance ascend to their normal level and the ratio of maintenance hours to production goes up. Management could then say: "Why was your performance so good last month?" But this is irrelevant since the change was exclusively due to the character of a workload imposed from outside.

Jobs which are not within the true scope of maintenance of *existing* plant and equipment but rather dealing with *new* plant, therefore, should be kept apart in controls and reports. This group includes: capital or work related to capital investments; plant development or expansion; constructions or installations; modifications, alterations, adaptations, etc., and transfers or dismantling. In practical terms, however, whether you pick up a screwdriver to repair an old machine or to install a new one, you are a mechanic and belong to the maintenance department.

The best way to describe the functions of the maintenance department is to group them as shown in Table IV. Referring to the five columns in this table, the following comments can be made:

(*a*) The grouping-together of complete *plant and equipment inventory* is as if we were to say, "Look, this is what we have to maintain."

(*b*) *Management and administrative duties* which have to be assigned to the maintenance staff to cover the life of plant at its various stages.

(*c*) The *type of services* required by the plant: figuratively, a wrench, an oil-can or an avometer.

(*d*) The *degree of planning* involved, such as emergencies, periodically repeated cycles or "do-it-any-odd-time" jobs.

Table IV. The functions of maintenance.

To maintain the following groups of plant inventory	(b) To manage the plant throughout its life	(c) To perform the types of jobs required	(d) To plan and schedule the workload	(e) To work towards specified objectives
ildings and grounds	Revision of plant	Adjustments	Emergency repairs	Expansion of plant
	Advise on new plant	Repairs	Unpredictable and irregular jobs	Renewal of equipment, reconditioning and replacement
	Examination upon arrival	Reconditioning of parts		
chanical equipment	Plan and install	Manufacture of parts	Planned plant expansion	
	Running in	Overhauls		Counteracting deterioration
	Instruction of operators		Unplanned alterations	
	Servicing during operation	Lubrication		
ctrical equipment	Testing and research	Inspection	Shutdown overhauls	Minimising wear and tear
	Disposal and salvage	Replacement of parts	Scheduled overhauls of machines	Preventing failures
	Administrative	Alterations		
fety and security vices	Keep manuals	Installation	Routine and regular service	Detecting faults
	Manage spares stock	Transfers		Improving performance
	Specify and plan the service schedules	Dismantling	Permanent assignments	Repairing breakdowns
fice equipment d furniture	Keep records	Testing and research	Patrolling supervision	Repairing effects of accidents
	Plan for replacement			
	Work planning and allocation			

(*e*) The *reasons for doing* the job, *e.g.* repair of breakdowns, prevention of failures, detection of faults, etc.

Going down each column we realise what people may mean when they ask the question, "What are your department's functions?" In attempting to reply without using Table IV as a guide we tend to jump from one group to another. Going through Table IV, check whether the groups listed also exist in your own company. Each group leads to certain organisational measures that have to be taken in order to make the system a thorough one. Let us now take a closer look at each group.

(*a*) The first group is helpful in setting up the recording system and for assessing the amount of work in each group. As a result planning becomes easier. The listing of plant in this form can also be used to survey the condition of all assets periodically.

(*b*) The second group can lead to better allocation of management and administrative duties among the maintenance staff. By detailing this group we can ensure that there are no omissions in the assignment of duties.

(*c*) The third group gives an indication of how to subdivide the department into teams. Some types of jobs will form natural groups as indicated in Table IV. The amount of work in each group may also indicate the relative strength in manpower.

(*d*) The fourth group distinguishes between jobs according to the degree to which they can be scheduled. Some jobs do not require scheduling and in others no scheduling is possible. The implication of this is fully discussed under "How to deal with the workload" in Chapter 6. Evidently, those who complain about their inability to plan and schedule are faced with too many emergencies and unpredictable jobs. Management sometimes adds to the confusion by requesting unplanned alterations as if they were emergencies instead of establishing a plant expansion programme.

(*e*) This group clarifies our objectives in providing the maintenance service. The emphasis may shift from "minimising wear and tear" to "improving performance." In some cases prevention of failures is a foremost requirement, as in power-generating plant or aircraft. "Counteracting deterioration" may be a temporary aim to extend the life of a plant until a complete replacement is available. Distinctions between the reasons for doing certain work are often revealing, particularly during discussions relating to budgeting.

To define our specific case we have to review these groups and decide which of the items reflect our own circumstances. When an answer to this question is established, disagreements about budgets, manpower, tools and controls will gradually diminish.

What is expected of maintenance depends on who does the expecting. Answers to this will never coincide. It is the equivalent of asking "What does a man expect of a pretty girl?" The answer will vary according to whether the man is young or old, married or single, a film-producer or simply the girl's father. Similarly, every person within the enterprise will have his own opinion about maintenance. The example is not meant to be facetious. Just as a pretty girl cannot satisfy everybody's expectation so the maintenance department cannot please everybody. Everyone will have his own expectations of maintenance. A way must be found to satisfy conflicting requirements such as for high quality of service against budget economy, for frequent servicing against minimum interruption of production, and for extended life of equipment against heeding the onset of obsolescence.

Chapter 4

Planning the system

Poorly organised maintenance departments are not hard to find. In most cases good reasons are given for the existing state of affairs. Here are some of the opinions taken from a rich collection acquired over the years:

"Our machines are so old that no amount of maintenance will help them. Anyway, we are planning to replace them."
"This equipment is brand new and highly automatic, all it needs is a few drops of oil in the right places."
"We've had no trouble for many years just as we are, these are sturdy machines so why worry?"
"Our operators have instructions to take good care of their machines and it works!"
"We've tried to get experienced men for years, but in this area, they're simply not available."
"We rely on outside contractors. They're right next door, you know!"
"Our foreman knows all the machines like the back of his hand, we don't need a system to tell us what's wrong."
"We have a bunch of good lads, they work hard and would resent being tied down by procedures."
"We get all our figures from accounting."

It is overlooked that in instances such as these the company is wholly at the mercy of events. No provisions are made, no precautions are taken, no direct action is initiated unless the roof caves in, and when it does the situation is out of control. Above all, there are no figures to indicate what the present lack of a system is costing.

In other instances a semblance of a system seems to prevail. There are procedures, accepted functions, estimating of costs and apparent controls. Yet on a closer look, maintenance is found to be continually on the defensive, in a rush, or being reorganised. It is as if this function had been superimposed carelessly on the rest of the structure. Management takes little interest in this activity except to blame maintenance for every conceivable mishap and, since this attitude is infectious, everybody else does the same.

Sometimes problems arise as a result of an overwhelming amount of procedures and paperwork and in other cases from a clash of personalities. Let us not lose our confidence in human nature, however, for in the end common sense usually prevails. What is needed in many cases is an approach which considers all the ramifications of the maintenance function within one integrated cycle. This is the approach described and advocated in the following sections.

THE "SYSTEMATIC" APPROACH

All the activities that take place in and around the function of maintenance fall into one of the following categories:

management techniques;
clerical procedures;
technological practice;
personnel management;
performance controls.

It is a combination of these elements that will give us the most suitable system for our needs. Whether we realise it or not, they do exist even in the smallest maintenance groups, although the functions may be neither separately identified, nor filled by full-time position-holders.

The system itself consists of a number of stages which together complete a "loop of action" (*see* Fig. 10). When properly implemented this loop will be self-regulating. One basic aim of this book is to show the advantages to be gained by applying this approach to the establishment of maintenance systems or to the analysing of existing ones. There is probably no maintenance department where all these aspects are fully covered. In fact, no situation is ever perfect and there is always room for improvement. Operating efficiency depends on employing the available measures and techniques to their greatest extent, *i.e.* each contributing to the system more than what is basically required of it. We know, from practical experience, that techniques are usually not exploited to their fullest potential.

Take, for instance, job timing and estimating. These should not be considered solely for the purpose of introducing incentives. They are equally important for scheduling the work of tradesmen, for cost controls and for performance measurement. The potential uses of these measures are many and they will be detailed in later chapters.

Similarly, personnel procedures are poorly exploited. Name, occupation, craft grade and serial number are the basic details with which we concern ourselves, while an inventory of skills, potential for promotion, seniority, craft up-grading are neither systematised nor followed through in the majority of cases.

The "loop of action" tells us the best way to develop a suitable maintenance service in a systematic way. By going through the sequence we will establish:

what is to be done, and *why*;
how the work is to be done, and *when*;
where the job is and *who* is to do it.

Since it is simple to check, at any time, conformity of our system with the proposed sequence, we can also absorb new developments without disrupting the operation. Such changes are continually being imposed on industry and we cannot but adapt ourselves to them. Such changes are also dictated by increased workloads.

Let us take, for example, the introduction of a computerised lubrication and in-

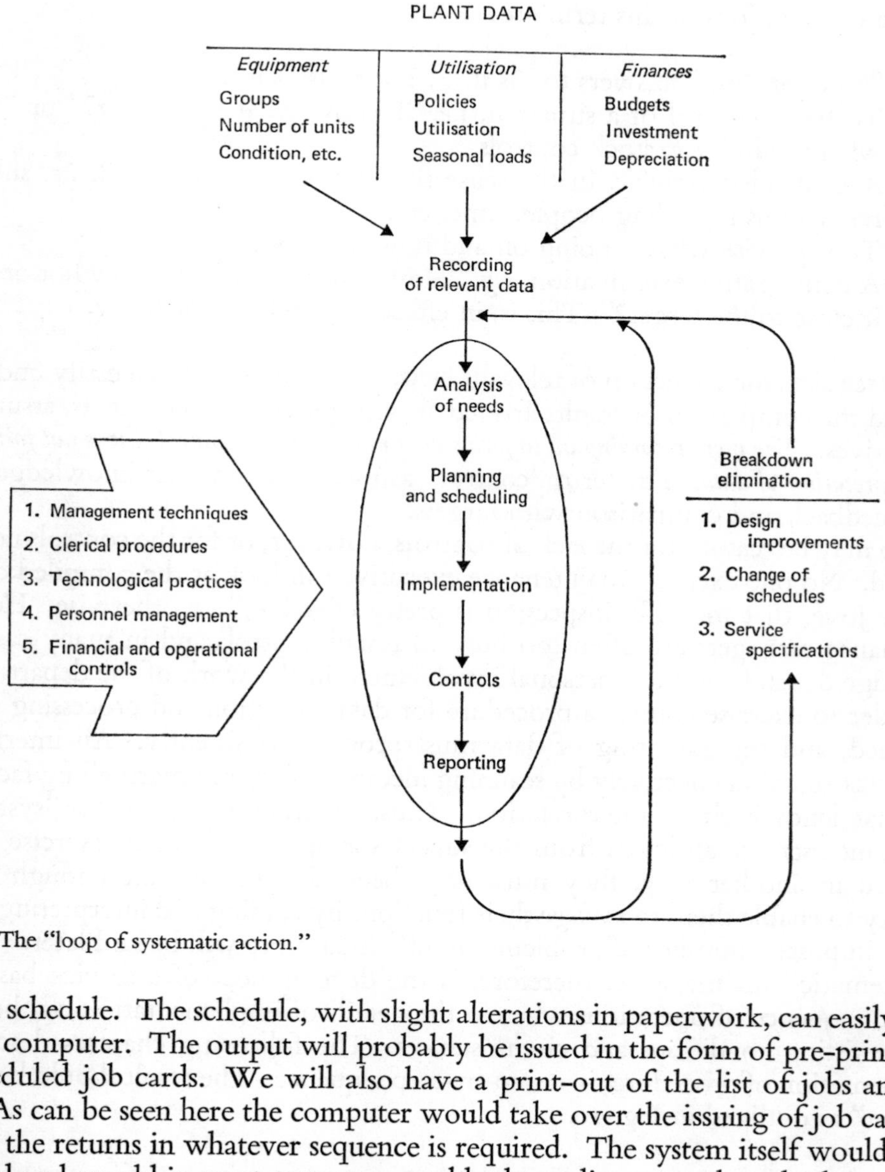

Fig. 10. The "loop of systematic action."

spection schedule. The schedule, with slight alterations in paperwork, can easily be fed into the computer. The output will probably be issued in the form of pre-printed and pre-scheduled job cards. We will also have a print-out of the list of jobs and their status. As can be seen here the computer would take over the issuing of job cards and tabulate the returns in whatever sequence is required. The system itself would not be disturbed and would in most respects resemble the earlier manual method.

Another change that may occur is the decentralising of some activities. When such moves take place, a clearly defined system will be helpful in indicating which of the jobs should be decentralised, how others are affected and how the procedures can be operated effectively. Since each phase is distinct and well defined it is convenient to deal with them one by one.

The most frequently overlooked stage in the "loop of action" sequence is that of *control*. In fact, the meaning of "control" is often not fully understood. Some of the common connotations of this term are:

(*a*) "Yes" or "No" answers to "Is it. . . ?" or "Isn't it. . . ?"
(*b*) To be in control of a situation, *i.e.* "I know what I'm doing?" or "I know where I am," *e.g.* stock controls.
(*c*) A controlled activity, in the sense that it is not a free-for-all, *i.e.* subject to restrictions regarding people, time, etc.
(*d*) To supervise what is going on and how it is done.
(*e*) A comparative examination, *e.g.* "How much, how little, how hot or cold, is it close to the target?" This is, in effect, a monitoring activity.

Control requires the collection of relevant facts, their analysis into an easily understood form and the comparison of results arrived at with previously set targets, assumptions or objectives. *The mere recording on to forms does not constitute control if it is not followed by the interpretation of data.* The term "control" as used here indicates knowledge arising out of feedback and comparison with targets.

There may be reasons for the lack of controls, however, or for the controls not being exercised. No one except a maintenance executive can look at the compiled data and say "By Jove, that monthly inspection is pretty effective," or otherwise. However, maintenance managers are often too busy to install controls and in many cases their knowledge depends on their personal involvement in the work of the department.

In order to exercise control a procedure for data collection and processing must be established, and the gathering of data must proceed consistently. All intermediary stages must function effectively by scanning incoming data and marshalling facts into a form that lends itself to interpretation. Thus, in order to operate a "systematic" maintenance service all levels from the supervisor upwards have to exercise control. Expressed in another way, they must have been able to delegate enough of their authority to enable them to *manage* their functions by reading and interpreting figures. This again poses a number of problems about which more will be said later.

"Systematic" maintenance, therefore, is the development of a service based on a specified sequence of operations wherein the details of implementation can be chosen to suit existing conditions and circumstances. The following chapters describe the implementation of procedures in a step-by-step sequence. The result should be a "loop of action" as outlined in Fig. 10.

THE FOUR SYSTEMS*

There are four distinct ways in which service can be set up, depending on a *rising degree of complexity* of the system and the way in which details are specified. To simplify the description we distinguish between the systems according to the overall approach to maintenance.

1. *Breakdown maintenance*—"Don't do a thing until you see smoke rising."
2. *Routine maintenance*—Simple service in a regular way, or "as much as we can afford."
3. *Planned maintenance*—according to the needs of the equipment and its utilisation, specified along an annual time-scale.
4. *Preventive maintenance*—"We'll do anything to prevent breakdowns."

The above terms seem to have gained some acceptance, although they are often misused. Other terms appear from time to time. For example, the term "productive maintenance" means an effort to set up the function on a planned and measured production pattern. The output relates to the number of servicing tasks completed, *e.g.* lubrication, inspection, overhaul, etc. The term itself has a kind of "promotional" value and, in fact, was originally used by a motor manufacturing firm in the U.S.A. to denote that motors of any size could be sent to them for reconditioning. At the factory, work is done on an assembly-line basis which makes it economical and uniform in quality. As we can see, the term would not be suitable for the average plant.

"Corrective maintenance" is a term used mainly in the U.S.A. to denote efforts to minimise possible faults and defects by applying certain corrective measures. It is also an attempt to reduce the *need* for maintenance and thereby lower the total maintenance costs to the company. This approach relies to a great extent on feedback of technical information.

The emphasis in this approach is on obtaining full information of all breakdowns, their causes, their duration and the incurred costs of down-time. Having obtained this information, efforts are made to eliminate such breakdowns by (*a*) changing the process; (*b*) redesigning or improving a component that has failed; (*c*) rescheduling the maintenance service; and/or (*d*) altering operating instructions. The implementation of this concept is quite effective but it does not qualify as a system; it remains merely a concept.

The term "corrective maintenance" as used in the U.K. refers to services carried out to restore an item to an acceptable working condition. In this respect it is akin to the term "repair maintenance." Neither of these terms seems to convey an

* The terms used in this and subsequent chapters differ to a certain degree from the ones adopted by the British Standards Institution's B.S. 3811 of 1964. This can be explained by the author's intention to differentiate in a concrete way between operational systems in actual practice. This approach has been adopted rather than trying to fit these systems into a listing of preferred terms, backed up as they may be by a more widespread usage in the United Kingdom. At the time of going to press, B.S. 3811 is being revised in order to cover more recent developments.

indication of a systematic approach; both are, in fact, covered by the term "breakdown maintenance."

A common term used in the U.K. is that of "designing out." It denotes the effort to eliminate the necessity for maintenance by improving design details of plant. Specially encased bearings, friction-free movement and automatic lubrication are a few of the results stemming from this approach. For example, we could have expected designers of some plant to avoid the use of different metals or alloys which promote corrosion by acting as electrodes. Or, if this could not be avoided, to provide for protective coatings. This, however, has not been the case and cathodic protection, as an after-the-fact antidote, has still to be applied. Again, this cannot be defined in an operational system but is rather a long-term goal to reduce maintenance workload and costs and increase equipment availability.

Today, however, equipment manufacturers are faced with stiff commercial competition combined with systematic sales efforts. The user often benefits from the fact that easy maintainability and trouble-free running often feature as selling points in the advertising campaign. This does not mean that some other components of the equipment have not become more vulnerable. Since machinery is becoming increasingly complex, the interaction of sub-assemblies, or clusters of components having certain functions, greatly multiplies the chances of failure. This latter approach has led to "unit-replacement maintenance" where complete sub-assemblies are replaced with a stand-by unit. Such factors are taken into account in the reliability concept which attempts to balance the cost of more reliable, and therefore more expensive, components against their premature failure and the resulting losses. The probability of such failures, their frequency and their anticipated costs form the basis of this theory. The major part of this approach was formulated in connection with the U.S. space programme. However, this method is not as novel as the space-ship that we envisage goes with it. In the car-fleet business—whether buses, taxis or military vehicles—this has long been the accepted practice. When a vehicle comes in for overhaul, such parts as alternators and distributors or whole engines are replaced with reconditioned units taken off the shelf.

An American police garage has held competitions between repair crews to see how quickly complete engines could be replaced. For a time the record was held at about thirty-five minutes by a crew of three working on a Ford pick-up. This sort of result is feasible provided that interchangeability has been attained by deliberate effort.

Servicing of aircraft on the basis of replacements is, in addition, not a new concept. During the Second World War the R.A.F. succeeded in perfecting its ground service so as to minimise aircraft servicing time, thereby maximising availability for operations. Considering all the types and changes of models, this was no mean achievement.

We often wonder at the meaning of terms so widely used and so freely interpreted. What is meant by "running maintenance"? Is it the work done on vehicles to keep them running, or is it "maintenance on the run?" Or, again does it mean that the vehicle is serviced without recalling it from service? We can only assume that people who use this term know what it means.

The terms appearing at the beginning of this section were introduced in an attempt to define the main objectives of the service, or those parts of it where most of the effort has to be concentrated. Each one represents a system which *grows in complexity* as we progress from breakdown to preventive maintenance. *The systems are not mutually exclusive* within one particular enterprise; they run side by side, complementing each other. However, a certain plant unit can only appear in one system at a time. For example, pumps are serviced on a routine basis, boilers by a preventive service. Let us discuss each of the four systems in turn.

Breakdown maintenance

This approach is also referred to as "repair" maintenance but it is not in the true sense a system at all. We can find here none of the stages of our "loop of action." No service is carried out, with the exception of an occasional lubrication, unless a failure has occurred. Sometimes, no maintenance men are on call, and in metal-working shops, machine operators often repair the machines that they themselves use, although at times a mechanic or an electrician may be called in from subcontractors. No effort is made to find out the reasons for the breakdowns. There is no stock of spare parts and, of course, no budget and no records.

At first sight it looks very economical and perhaps for a short period it may well prove so, but management is getting no information on how much it costs to keep the plant running, there is no summary of time lost due to breakdowns and only occasionally does a shop foreman complain that he can do no more. When a repair bill has to be paid management shrugs it off and instructs the book-keeper/accountant to charge it to "miscellaneous" expenses.

Thus, all the trouble created by the lack of properly organised maintenance is "submerged," until breakdowns occur with ever-increasing frequency. In the absence of repairmen every job becomes an emergency. Who is to do the repair? From where do we get the parts? How do we pay for them? Who is to go to town to buy the parts? Everybody, from foreman to manager, becomes involved in overcoming the emergency. This again costs time and money and in the meantime the machine is standing idle. When finally an *unreliable job* is completed at *excessive expense*, we realise that nothing has been solved, and next time the same process is repeated. By ignoring the obvious remedy, irritations are piled on top of costs. A slight improvement may be achieved, though, by arranging for a contractor to be on call and by stocking some spares. Teaching machine operators to service their own machines may also help to reduce breakdowns.

Such situations may be warranted in small plants where, so the argument goes, there is no full-time work for a mechanic–electrician, nevertheless it may well prove dangerous. An example can here be cited where lack of maintenance contributed to the ruin of a business.

A small lens-polishing plant, employing twelve polishers, called for the assistance of an industrial consultant. The manager–owner said he was not making money and could no longer finance the purchase of raw materials. At the manager's request the consultant visited the factory where he found the shop in disarray with parts of machinery on the floor and empty cartons strewn all over the place and during a tour round the polishing benches he observed the manager was eyeing his workers with a good deal of suspicion.

From the questions and answers in the "office" the following story was reconstructed. Production was on a bargained piece-rate basis. Everything went well until a second-hand machine started to fail. At first the operator could only produce a limited amount of lenses which earned him less than by working on a day-work rate. So he was put on day rate. Then, out of solidarity, the rest of the workers decided to take turns on the defective machine. The manager opposed this, since he did not want his good polishers to waste their time on an unreliable machine. He insisted that three of the best operatives should work only on their own machines and the rest should use the defective one alternately. The workers complained that this showed favouritism and demanded the urgent repair of the broken-down machine. The manager tried to postpone the repair until enough money came in from his sales.

At this point another machine broke down and a few days later fist-fights broke out among the workers and the manager was threatened with violence. A repair mechanic was called in who advised the manager to buy two newer machines that were available at a bargain price. When the workers heard this, three of them pooled their resources, bought the machines and set up their own business. They hired the mechanic, after paying him commission on the machines, to build attachments that they had seen in a catalogue. The dissatisfied customers of their former employer then turned to them with their orders.

For all its shortcomings, there is one situation when breakdown maintenance is justified, not as a system, but as an approach to solve a certain problem. When down-time is very costly it is preferable to introduce periodic shutdowns, then run the equipment until it fails, or shows signs of failing, and shut down again. This approach can be adopted in some instances, provided that the quality of production does not deteriorate. In some cases it may be possible to wait for a failure to occur and then replace assemblies with stand-by units rather than attempt to repair them.

Two large motor-car manufacturers in the U.S.A. have different approaches to this type of maintenance. One uses an ample crew of patrolling repairmen to correct defects as they arise; the other relies on stand-by replacements after failures occur. The cost of keeping a wide range of spare units may, however, make the latter approach prohibitive for most companies.

Routine maintenance

The dictionary defines the word "routine" as "a procedure followed regularly," or "a habitual sequence." The term is used here in a similar sense. When we lay down a rule for performing a certain procedure in a predetermined manner, we are establishing a "routine." We may extend this definition to "a cyclic operation recurring periodically." Examples of routines instructions are:

Check all compressors first thing on Monday mornings.

Lubricate completely two machines daily (or weekly)—at this rate all machines will be attended to within a certain period, and then the cycle recommences.

Service certain machines every 1,000 hours/miles/tons.

On Monday mornings for four hours carry out inspection work, starting at a certain point and progressing as far as possible until lunch-time. Next week carry on from that point for four hours. Keep up the sequence until the cycle is completed.

On Mondays lathes will be serviced, on Tuesdays—presses, on Wednesdays—grinders, on Thursdays—millers, on Fridays—transport equipment.

All motors are to be inspected during the first week of every month, all starters in the second week, etc.

Thus routines are established by defining the amount of time available for work, the number of units in a given period, or the type of work to be performed at specific times. In every case *a pattern* is set which *can be repeated over and over again without the need for issuing any further instructions*. By estimating the time required for the jobs we can calculate the period required to complete a cycle, and thereby we arrive at the frequency of attention that a unit will get in a month or a year. Or, conversely, by defining the frequency of a service on which we insist, we can derive the total number of man-hours that the service will require.

The attractiveness of routine services lies in the following advantages:

(*a*) they are simple to establish and to follow;
(*b*) little or no clerical work is required;
(*c*) this type of service achieves a high degree of prevention, by intercepting developing faults.

It is a fact that many routine systems are called "preventive" because of their effectiveness, especially in instances where previously neglect used to prevail. Compared with a "breakdown" approach, every unit of plant will be serviced at least one or more times per year. On those occasions, defects which have been slowly developing will be discovered and eliminated. Similarly, a high percentage of failures, which may otherwise have developed due to total lack of attention and/or lubrication, will be avoided. Hence this system is often erroneously called "*preventive* maintenance." As discussed below, preventive maintenance does contain services at regular intervals as part of a more elaborate system.

A more advanced stage of this type of system calls for service instructions to be issued on a pre-printed schedule or checklist in order to be easily understood and followed. These can easily be checked for compliance and the recording of their performance can be made by a simple X in an appropriate column on the form. The yearly cost of such service in man-hours can be easily compiled by multiplying the job times by the number of machines serviced and by the yearly frequencies (*i.e.* time × frequency × no. of units).

Undoubtedly, for the time, money and effort invested in planning, implementation and control, this system yields outstandingly good results. The regular attention given to plant achieves a high degree of "prevention" and vastly prolongs plant life.

When a routine is defined we may not provide the service that the manufacturer of a machine has specified, and we may ignore information about breakdowns in preceding years. We will only be concerned with the fact that we can afford to service, say, twice a month or every two months and it will have to suffice. Without going into the details of a machine we establish an average service interval that has to be observed. Whether the main bearings require oiling every 100 hours and the cross-spindles once a week, we set servicing at twice-monthly intervals. We can also ignore the fact that some of the motors run one shift, others twenty-four hours a day or intermittently, and we call for a routine check of all motors on Mondays. A simple periodicity will overrule minor differences.

To simplify instructions, we issue a "blanket order" to service, rectify, adjust, lubricate or check all that may need attention. Checklists may or may not be prepared; job cards may or may not be required. Routine services usually include lubrication, checking, inspection or adjustments. Completion is recorded by ticking off the work done on collective scheduling sheets. Routine service has to be supported by a separate crew repairing defects that have been encountered, otherwise regularity of the routine service may be delayed while defects are being corrected. Such an arrangement should be considered as part of the setting up of a routine system.

Routine service, then, does not cater for particular needs, exacting requirements or fail-safe provisions. It emphasises easy-to-follow instructions which cover most plant units with a regular service, which is better by far than had been provided before and which can be done at a minimum cost.

Planned maintenance

In this type of service the emphasis is placed on the machines. What does the manufacturer prescribe? Is the unit utilised for two, or three, shifts per day? Is it working under a normal load? Are conditions as good as those envisaged by the manufacturer? Do we allow for extra attention owing to corrosion-inducing conditions? Are there any other factors that may detrimentally affect the equipment?

Instructions are more detailed in this system than in the case of routine maintenance and may call for differently timed servicing on the same unit. Thus, in an automatic furnace the roller chain may require monthly attention while the injection burners require weekly checking and the instrumentation a quarterly calibration. When the frequency of service for all items is established, the dates of annual services are set. This again is different from routine maintenance. Planned maintenance requires the work-load for the team to be planned in advance for every week. This entails both a planning effort, which may be considerable, a faithful implementation and, of course, recording.

Planned maintenance will take into consideration changes in conditions of use or increased wear of parts. As a rule inspections, replacement of parts and adjustments are included in the overall plan. During the planned service, detailed instructions are to be followed to reduce the chance of failure during the period extending to the next service. Unforeseen work is thereby greatly reduced.

This system provides as much attention as the equipment requires—to the best judgment and ability of the planner. The shorter the intervals and the more detailed the service, the better insurance we have against breakdowns. As more experience is available the periods may be "stretched" and some details of work omitted in order to achieve maximum economy. The initial list is often deliberately over-elaborate when new equipment is installed in order to allow service crews to acquaint themselves with all parts. It is later reviewed and reduced.

In order to obtain the greatest benefit from this system, and to operate it with maximum economy, planning should be thorough and recording must be followed up. Analysis of recorded data will assist in manpower scheduling and production planning by indicating servicing times. Past failures may point to inferior parts or materials that will have to be avoided in the future. In addition, analysis of the reasons for breakdowns may indicate the need for action concerning operator-training, materials used and the equipment itself. These advantages can only be achieved by correct recording and interpretation in which all concerned must play their part.

Preventive maintenance

This term applies only to systems which strive to reduce the likelihood of failures. Such is the case in the operation of aircraft, power-stations or critical installations for certain plants, such as steam in a laundry or forced ventilation in mines.

To achieve prevention of breakdowns, planned service is carried out with the explicit additional objective of *detecting weak points and ensuring perfect functioning by replacing parts* which could still be used were it not for the assurance that we require. Thus after every service a machine is "as good as new" and has a high degree of reliability. In addition to this, routine inspections are carried out in the interval between planned services. These are also directed towards eliminating possibilities of failure. Thus, with added costs, maximum reliability of operation is obtained.

Preventive maintenance occasionally employs statistical methods for determining life expectancies of parts and materials and it thereby establishes more accurate replacement periods. Measuring devices are also employed during check-ups in order to detect changes which might indicate deterioration, such as in electrical insulation, or deformation stresses exposed by X-ray. This phase has been termed "predictive maintenance."

Naturally, the cost of running this system is high and it should therefore be applied only in cases of absolute necessity. Both the planning of preventive maintenance and the replacement of parts before they actually fail make this an expensive system.

Eliminative maintenance is a term used to designate an approach that strives to minimise the necessity for maintenance. This has also been called "designing-out" the necessity for maintenance and it can best be applied on the drawing-board. Thus plant can be so designed as to require extremely little attention during its lifetime, *e.g.* permanent bearings. This idea is now being applied commercially on motor cars, where sometimes materials are employed that are outstandingly long-lived.

In fact, this objective requires closer examination of all components with a view to making them infallible or easily replaceable. Centrally placed networks for lubrication are also being installed which allow close supervision and require little attention. Yet another aspect which has received much attention lately is the installation of plant so as to ensure easy accessibility in cases where space is restricted.

Depending upon the item under consideration this may be the task of either the designer or the plant engineer who installs the equipment. As mechanisation increases co-operation between the user and the manufacturer becomes vital. In a modern production unit we may find several mechanical systems and electronic controls. Unless all components are reliable and trouble-free their combined probability of failure may produce a nightmare. This approach to maintenance, which endeavours to make equipment less prone to failures, although more expensive, is also justified by high investments on modern plant. It is more fully explored in a book on reliability.[3]

SETTING UP THE SERVICE

Having discussed the basic principles of systematic maintenance it will be clear to the reader that a haphazard collection of activities having unco-ordinated objectives will not deserve to be called "a system." As soon as we can establish which approach will serve us best for a certain group of equipment, our work becomes purposefully directed and all haphazard activity must cease.

It can be argued that systems impose a rigid strait-jacket régime on the maintenance crew. This may be true if initially there is a "work-as-it-comes" situation. In such a case any imposed order seems oppressive. However, why should we not compare it with the discipline imposed on production workers? All their work is rigidly prescribed in a step-by-step procedure which has to be followed with regard to both sequence and timing. Why, therefore, should a lubrication schedule, with a well-defined route or a work-order system be considered oppressive? How then, do we substitute a system for haphazard work?

The first rung in the ladder leading to perfection is the establishment of *objectives and policies* as discussed in Chapter 2. This will answer the questions: *what* is to be done, *why* and *how much*?

The second rung is the definition of a *team organisation* that will allow us to work towards the objectives, as discussed in Chapter 3. This defines *who* is to do the job and *where*.

The next step is the *choice of a system* that will best serve our purpose, as described earlier in this chapter. Again, it must be emphasised that these systems are not mutually exclusive within an enterprise; however, we cannot use two systems for maintaining one machine. We have to choose the systems that will apply to the various parts of our enterprise, be it individual machines, installations or groups of process machinery.

The following rung is the *preparation of procedures* that will operate the system. To this end we need clerical work and this is discussed in the following chapter.

As we proceed to the ideal system we have to perfect the implementation of work, the planning, scheduling and supervision of the service and finally to define ways of exercising control. This will result from proper feedback. With each phase, work becomes more orderly, better defined and less capricious. The benefits of this transformation can be shown in practical terms and many of the changes can be quantified as we shall see in the chapter dealing with controls. Our basic purpose now is to be able to assess what we are doing, to direct our efforts consciously towards our objectives and to relate the effects of our efforts to the expenditure.

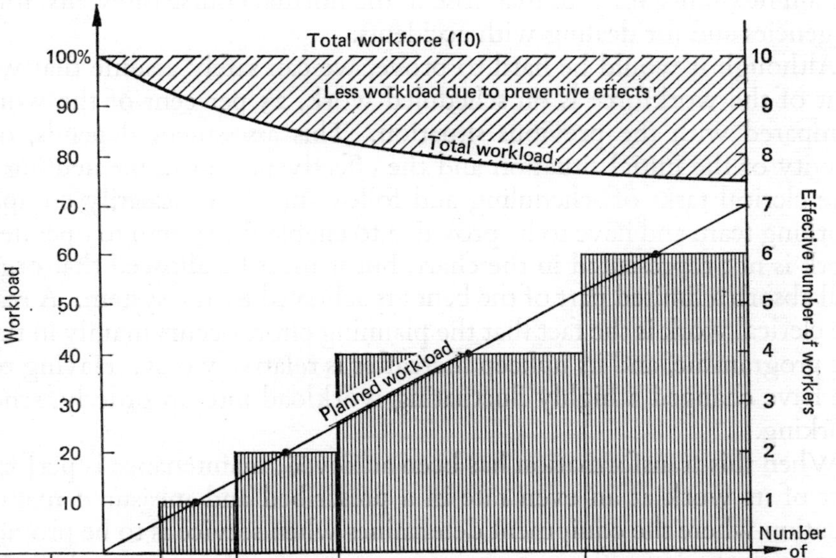

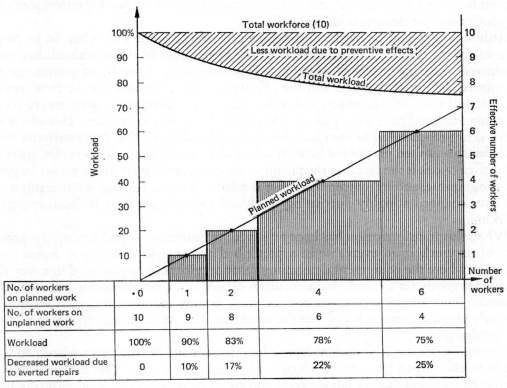

No. of workers on planned work	0	1	2	4	6
No. of workers on unplanned work	10	9	8	6	4
Workload	100%	90%	83%	78%	75%
Decreased workload due to averted repairs	0	10%	17%	22%	25%

Fig. 11. The effect of introducing planned maintenance on the total workload of a ten-man team.

A relatively simple chart can show what happens as we proceed gradually from the rushed activity of disorganised work to scheduled, repetitive service (*see* Fig. 11).

Let us assume that at the outset we have a ten-man team all working on jobs as they arise. This represents 100 per cent of the workload. As soon as one man is assigned to a regular load in the form of a lubrication or inspection schedule, he succeeds in averting jobs which would inevitably have arisen through poor lubrication or lack of attention. Therefore the rest of the team have a workload which is reduced compared with what would normally have occurred.

Another reason for this effect is the improved use of working time. Whereas irregular work involves a high percentage of wasted time in the form of waiting between jobs, receiving instructions, etc., scheduled work is mostly productive, except for the walking between one service point and the next. Thus many more points are attended to than would otherwise have been possible. This factor alone accounts for a great increase in productivity.

When two workers are on scheduled work a further reduction in the workload results. This progression is assumed to proceed along the curve in the chart. Gradually this effect diminishes until approximately 70 per cent of the staff is on scheduled work. Beyond this limit it is difficult to schedule. Every maintenance function requires a certain flexibility for jobs that arise in the normal course of events, for unforeseen contingencies and for dealing with accidents.

Although it would be hard to prove, we can safely assume that when 60 to 70 per cent of the workforce is on scheduled work, 25 per cent of the workload disappears compared with the previous situation. This assessment depends, of course, on the gravity of the initial situation and the effectiveness in implementing the new service. The clerical tasks of scheduling and follow-up are necessarily complementary to the working team and have to be provided to enable the system to operate. This additional work is not represented in the chart, but it must be allowed that organisational effort will absorb a limited part of the benefits achieved by the system. A reason for ignoring the clerical work is the fact that the planning effort occurs mainly in the earlier stages of the programme, and its upkeep thereafter is relatively easy. Having reached this stage, we have changed a highly fluctuating workload into an orderly, scheduled system of working.

When this transformation has been achieved, maintenance is performing the greater part of its work at an even rate in a prescribed and measured manner. Instead of a situation where the equipment determines when service is to be provided in a way that keeps everybody on the run, maintenance is carried out by *our initiative* and at *our convenience*. This allows further short-cuts, such as area maintenance, combination of services, planned overhauls and regular replacements. Maintenance then is no longer on the defensive.

In recent years the aspiration to have a high percentage of work on a scheduled basis has been challenged by a more selective approach to services. This can probably be attributed to steeply rising labour costs and to an increase in the reliability of equipment during a *deliberately pre-calculated lifetime*. As a result there has been a shift away from carefully detailed services to be performed at prescribed time intervals. Some advance systems are using statistical methods for predicting the likelihood of failures and others use them for diagnostic sampling. Thus shop floor know-how may be giving way to quantitative management techniques. For most of us such solutions are not yet practicable.

In summarising this chapter it will be apparent to the reader that the philosophies relating to the provision of maintenance services are as yet by no means unified. This is understandable since this branch of technology is still in the process of being formulated. Some people advocate reducing the *need* for maintenance; others assign various degrees of importance to the *frequency*, *speed* and *thoroughness* of the service.

In recent years some authorities have had second thoughts on preventive maintenance although others view this approach as the pinnacle of desirability. (*See* "Is preventive maintenance worthwhile?")[6] More recently the "don't-bother-bother-until-bother-bothers-you" school of thought has been reinforced by a principle defined as "condition-based maintenance" which, in conjunction with routine maintenance, is carried out as dictated by the apparent needs of the equipment (Ref. (E), November 1973). There is, however, no conflict between this approach and that implied by a routine or a planned inspection whose purpose it is to discover whether anything needs to be done.

Since maintenance is an axiom of ownership it is inevitable that the approach we eventually adopt will be a compromise between the alternative systems, whatever their official names, and our own preference. The system to be devised is *to prevent, reduce or eliminate failures, detect and diagnose faults and repair or correct the effects of use.* All this is to be done in the most economical manner in the given circumstances. The practical steps that will allow us to set up a system to meet these requirements are described in the following chapters.

Chapter 5

Paperwork

In modern industry there is a widespread tendency to underestimate the importance of clerical work. The clerical department is considered to be on the bottom rung of the ladder—even lower than maintenance—and the majority of managers and accountants regard it as a burden. Nevertheless it is a fact that most actions are initiated as a result of a clerical operation and there is usually no activity of consequence that is not recorded in one way or another.

Paperwork performs the following important tasks. It:

initiates action, by identifying the job and the date;
collects data, by describing the action taken;
controls operations, by recording the input of resources.

These three tasks are essentially linked and interdependent. As a rule, information accumulates following this sequence while a form passes from stage to stage. Unless a slip of paper accomplishes one or more of the above tasks, it is useless. It will be helpful to remember that certain written details serve as *input* data, others as *output* data and some as both, *e.g.* date of service. Again, unless a detail belongs to either of the above-mentioned classes, there is little point in recording it.

We tend to associate paperwork with the issue of job cards or service instructions and recording of work done. Its potential as a source of control data is often neglected and the forms remain unexploited in the files. The reason for this state of affairs may be ascribed to either too much or too little data having been collected or the data appear in such poor form that as soon as the action is completed it becomes useless. To avoid this we must seek to record selectively and in such a way as to allow easy retrieval and interpretation. By taking this action we can ensure:

the *relevance* of data,
their *accuracy*, and
their *retrievability*.

It follows that to avoid burdening staff with unnecessary work only relevant data should be recorded. Certain measures in the design of paperwork can be taken to ensure accuracy and easy retrievability. Filing cabinets are often full of used forms that are a direct contradiction to the above principles, piling up vast volumes of useless paper. In fact, the more voluminous the files the less useful they will prove to be. The information written on forms is frequently in excess of what is truly required, and often it may have been recorded in such a way as to foil any future attempt at interpretation.

Let us now consider the operational aspects of paperwork, namely the forms themselves. Written communications, though more time-consuming than verbal ones, offer some distinct advantages, provided they are well carried out. Forms allow us to divide activities and information into groups and to arrange them in a meaningful sequence. Thus we are able to deal with recurring actions both at the input and output ends in prescribed ways without straying from the issue, by using the same forms. Instruction sheets and checklists are examples of input data. Records in terms of hours worked and the consumption of materials on a given job are output data. Work requests can be grouped into mechanical jobs, electrical jobs or others. Equipment cards can be classified by types of equipment or by their location.

A form is also a guide, serving both the writer and the recipient, to the kind of information to be recorded and interpreted. The repetitive nature of a good procedure is habit-forming and thereby tends to minimise the time required by both writer and recipient to deal with it. Finally, a form carries the information unaltered and undistorted for as long as we wish and it will allow any kind of information to be written on it without argument. It will also allow us repeated reference to it without balking. These are some of the advantages of recorded information over verbal communication.

Consider these points in contrast to the situation that often exists in industry. Sometimes a certain kind of communication is so "very rare" that a special form is not required. Information, instructions, queries or notices go on slips of papers conveniently called "memos." Replies, if necessary, also come in similar form. A lot depends on the personalities of the writer and of the recipient and if after a certain time has elapsed we refer to these notes they may prove impossible to interpret. Interpretation by a third party may prove difficult and storage and retrieval almost impossible.

Perhaps the best feature of a good recording system is the fact that when a supervisor goes on holiday or becomes ill everything *does not* go with him. This argument must not be interpreted as advocating maximum clerical effort to prepare for all contingencies. Clerical procedures must be such as to *make the most useful data easily available without overburdening the department.*

A bad recording system is as much a curse as a good one is a blessing. To ensure a good system, it is necessary to revise procedures occasionally so as to keep up with new developments and organisational changes. Failure to do this is a pitfall that claims many victims.

Paperwork has to be tailored to the needs of a company. Rarely can we adopt a procedure from an outside source without slight changes. Some "packaged deals" are helpful, such as the "Keysort," "Visirecord," "Kardex" and "Kalamazoo" systems, but they have to be taken on with some reservations. The ready-made cards may be intended for a variety of applications, some of them different from ours or having different objectives. The situation as to scale of operation, available clerical and maintenance workforce may be different. Let us not arbitrarily commit both time and

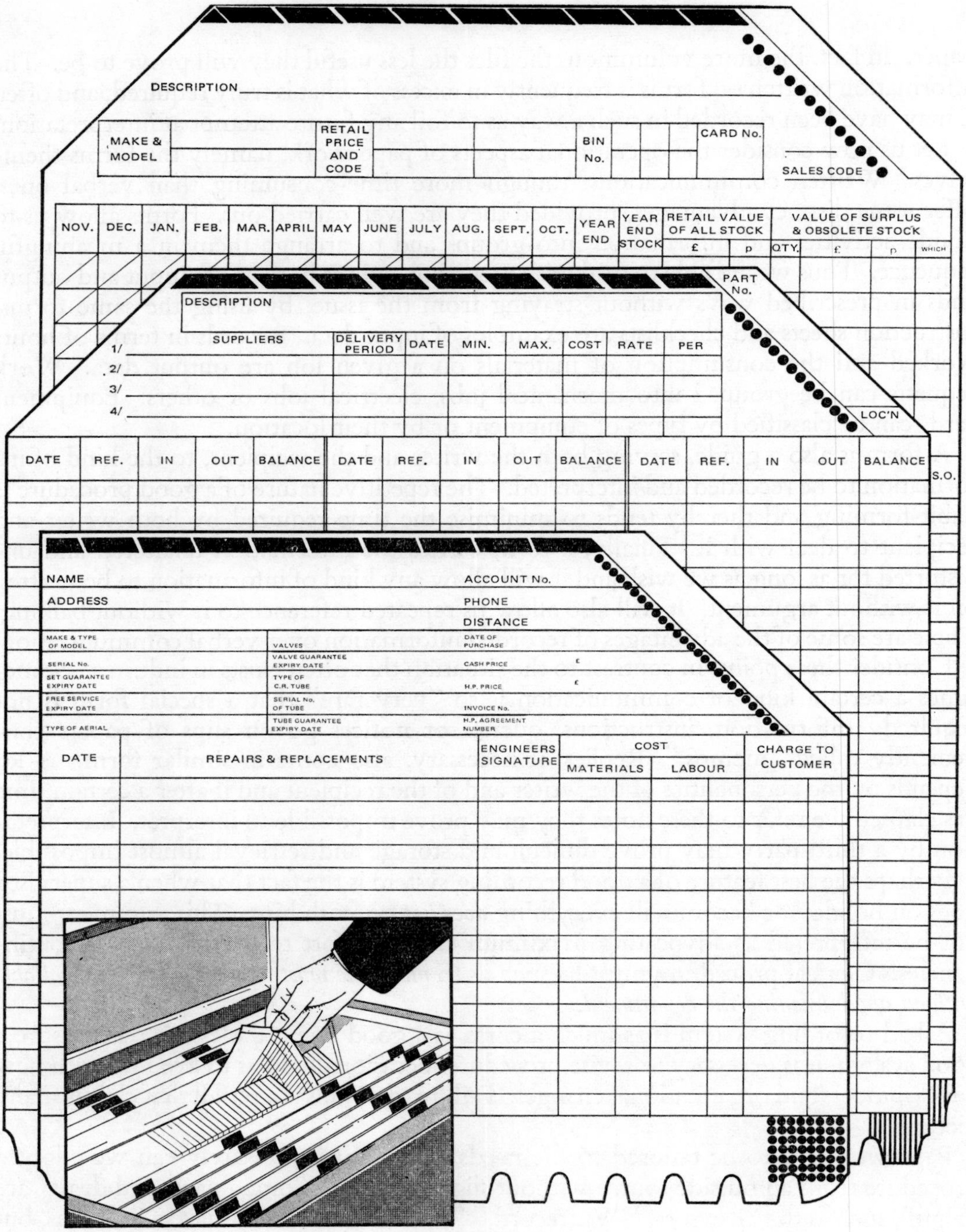

[*Courtesy Lamson Paragon Ltd.*

Fig. 12. This visible-card system can be adapted for scheduling and recording maintenance services. Each unit of equipment will have a history card, a service scheduling card and any number of instruction cards for mechanical, electrical and other services. These are taken out of the file during the week that service is due, and subsequently the service is recorded and the card replaced.

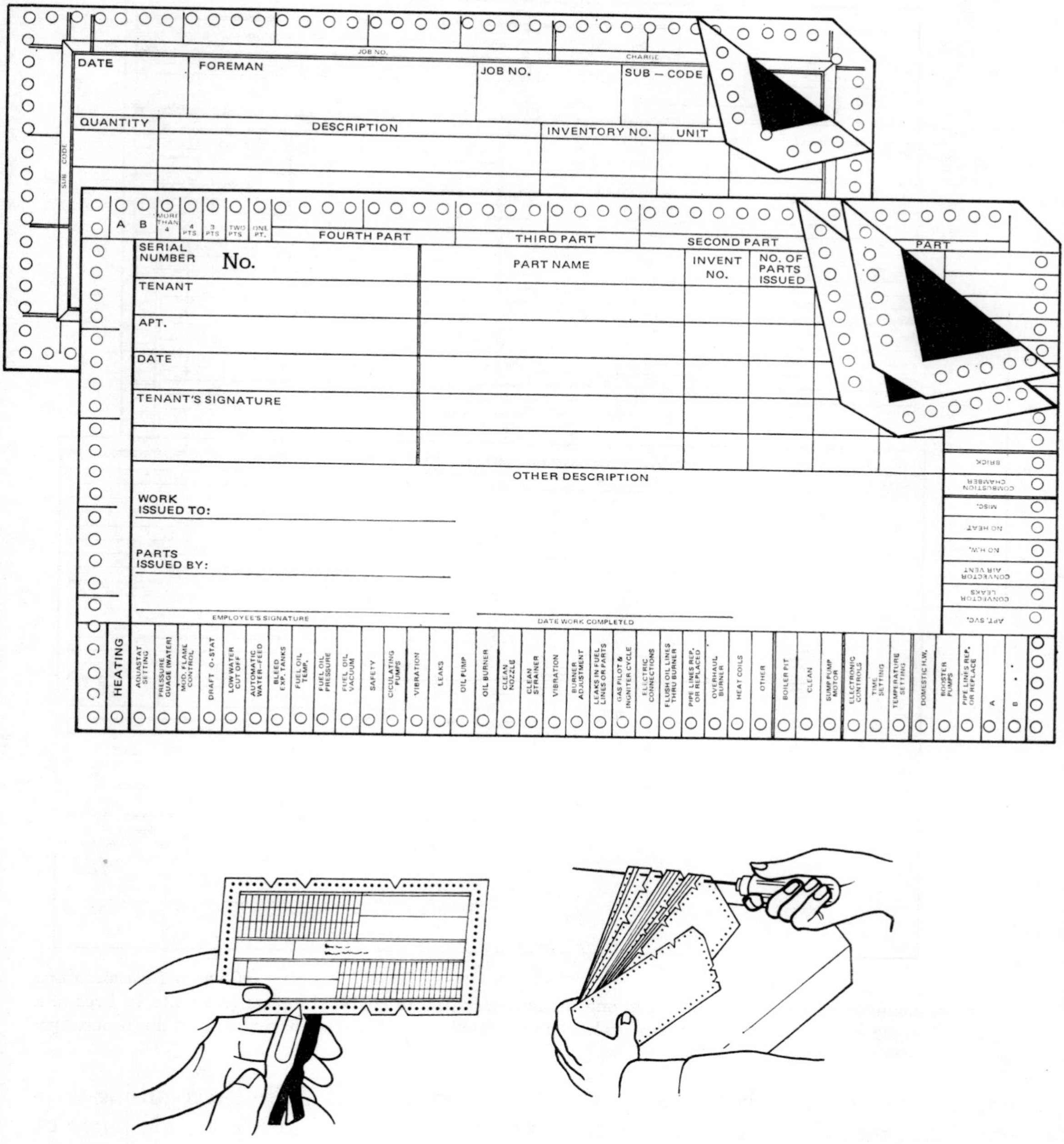

[*Courtesy Litton Business Systems Ltd.*

Fig. 13. The principle by which work orders are classified into a vast number of classifications, *e.g.* by trades, by departments, by types of machines, etc. It provides both the working forms and the permanent record. Pre-printing of classifications which often recur is also helpful. The notching is done when the job is completed. The "Keysort" system is illustrated above.

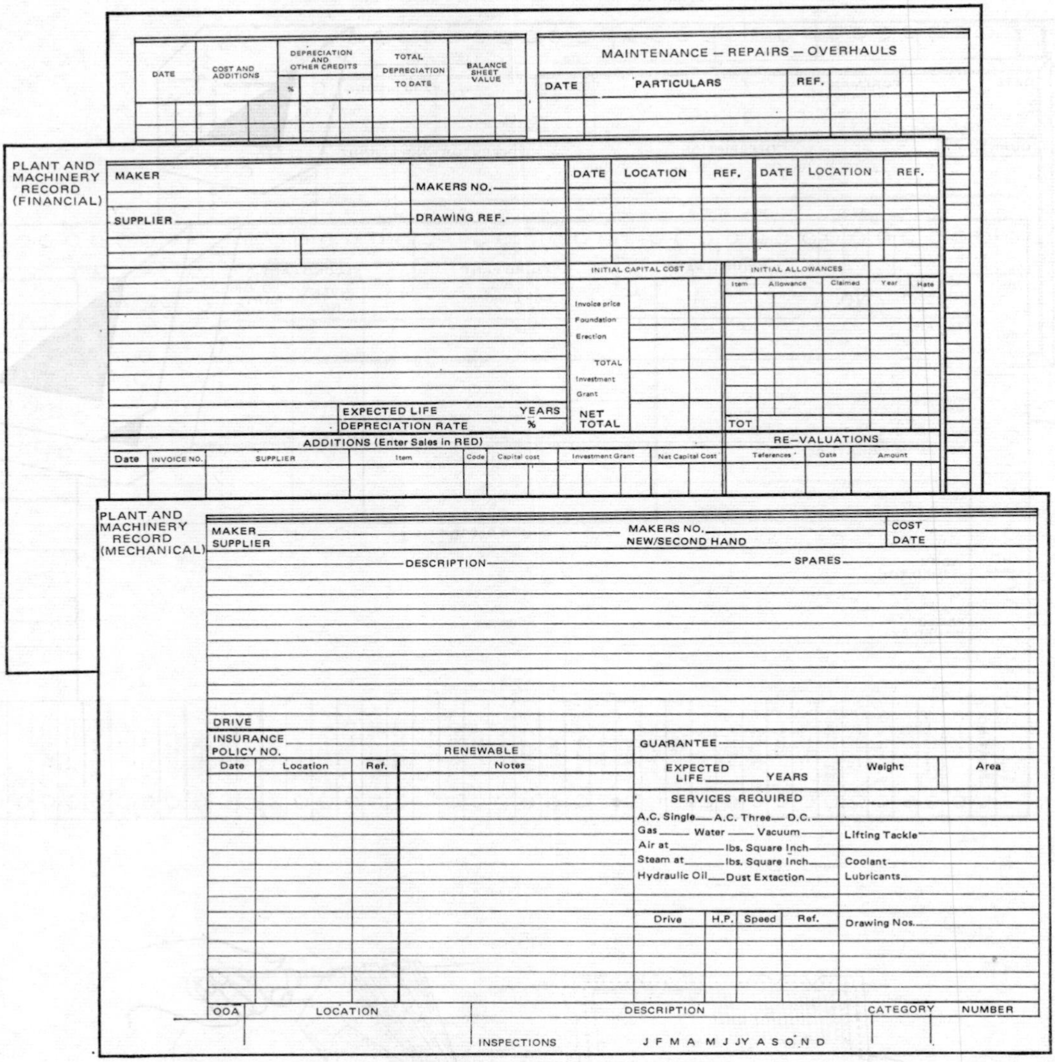

[*Courtesy Kalamazoo Ltd.*

Fig. 14. Examples of asset history cards, giving the necessary data to enable complete records to be kept of a company's equipment. The original documents have a perforated edge which is a characteristic of this proprietary system.

effort to such systems only because we get them off the shelf. While recognising their excellence they do cater for more than one need in order to satisfy a wide range of requirements. Assuming that most maintenance staffs in the U.S.A. and the U.K. number less than thirty, over-elaboration of recording procedures should be avoided lest we become saddled with unwanted rigidity and/or profusion of detail and/or duplication. Nevertheless, much inspiration can be drawn from these procedures and they are useful when the scale of operation finally demands that one of them should be adopted. Some examples of job cards that can help us choose a type best suited to our requirements are shown in Figs. 12–16.

Figure 12 shows a three-card system which provides complete machine data and maintenance records. It is a simple, flexible and effective method of planning, scheduling, communicating and controlling preventive maintenance. Figure 13 shows a

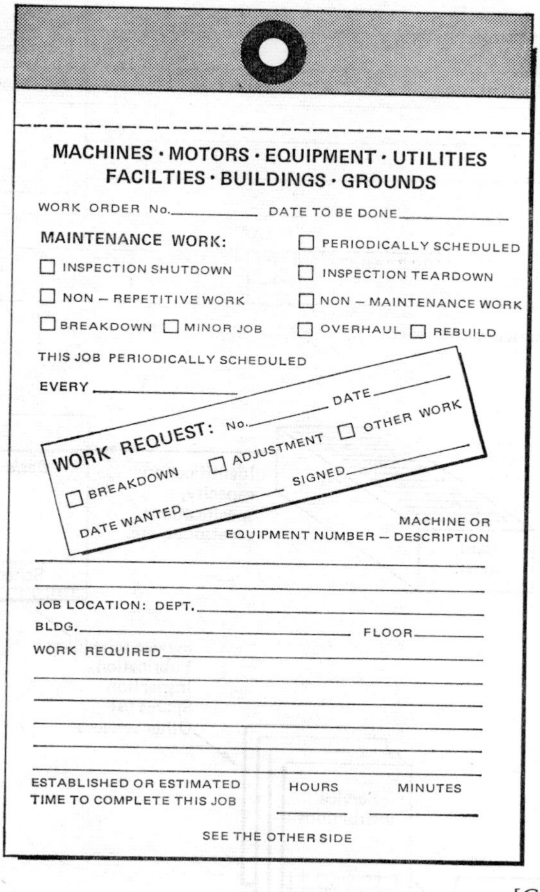

MACHINES · MOTORS · EQUIPMENT · UTILITIES
FACILTIES · BUILDINGS · GROUNDS

WORK ORDER No.________ DATE TO BE DONE________

MAINTENANCE WORK: ☐ PERIODICALLY SCHEDULED

☐ INSPECTION SHUTDOWN ☐ INSPECTION TEARDOWN

☐ NON – REPETITIVE WORK ☐ NON – MAINTENANCE WORK

☐ BREAKDOWN ☐ MINOR JOB ☐ OVERHAUL ☐ REBUILD

THIS JOB PERIODICALLY SCHEDULED

EVERY________

WORK REQUEST: No.________ DATE________

☐ BREAKDOWN ☐ ADJUSTMENT ☐ OTHER WORK

DATE WANTED________ SIGNED________

MACHINE OR EQUIPMENT NUMBER – DESCRIPTION

JOB LOCATION: DEPT.________

BLDG.________ FLOOR________

WORK REQUIRED________

ESTABLISHED OR ESTIMATED TIME TO COMPLETE THIS JOB HOURS MINUTES

SEE THE OTHER SIDE

[*Courtesy The Minute Company*

Fig. 15. A card for attachment to machinery in need of repair or maintenance.

system suitable both for detailed scheduling of jobs by means of equipment record cards as well as for separate job cards that are to be classified into groups for data-collecting purposes. An asset history card is shown in Fig. 14 while Fig. 15 is a card for attaching to broken-down machinery. For subsequent processing by computer work orders can also be issued partly filled in by a computer programme as is shown in Fig. 16.

It is ironical that good recording is most needed where maintenance is at its worst. But in the absence of recording how could management even suspect how bad a

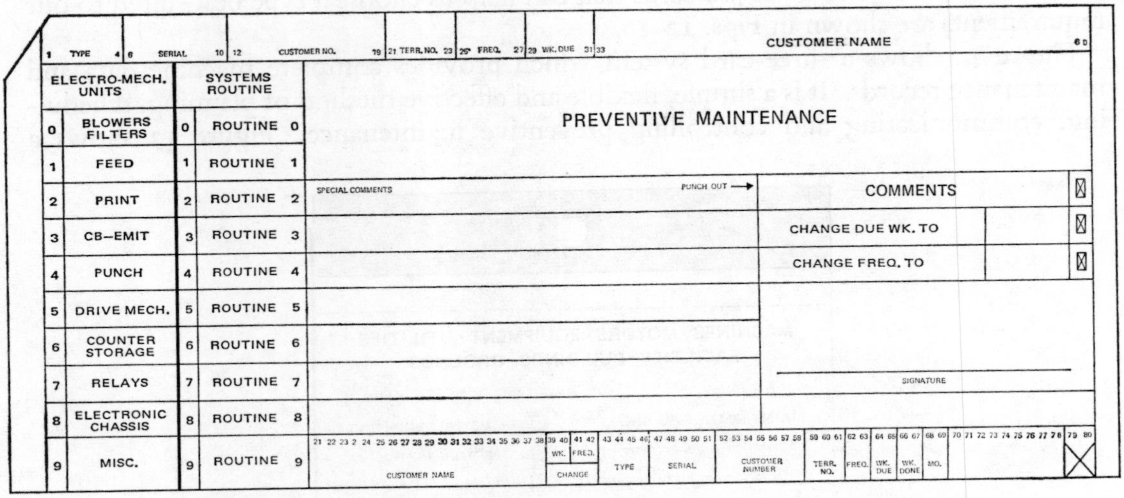

TYPE
SERIAL
CUSTOMER NO.
TERR. NO.
FREQ.
WK. DUE
CUSTOMER NAME
ELECTRO-MECH. UNITS
SYSTEMS ROUTINE
0 BLOWERS FILTERS
1 FEED
2 PRINT
3 CB–EMIT
4 PUNCH
5 DRIVE MECH.
6 COUNTER STORAGE
7 RELAYS
8 ELECTRONIC CHASSIS
9 MISC.
0 ROUTINE 0
1 ROUTINE 1
2 ROUTINE 2
3 ROUTINE 3
4 ROUTINE 4
5 ROUTINE 5
6 ROUTINE 6
7 ROUTINE 7
8 ROUTINE 8
9 ROUTINE 9
PREVENTIVE MAINTENANCE
SPECIAL COMMENTS
PUNCH OUT
COMMENTS
CHANGE DUE WK. TO
CHANGE FREQ. TO
SIGNATURE
CUSTOMER NAME
WK.
FREQ.
CHANGE
TYPE
SERIAL
CUSTOMER NUMBER
TERR. NO.
FREQ.
WK. DUE
WK. DONE
MO.

Fig. 16. Preventive maintenance work card.

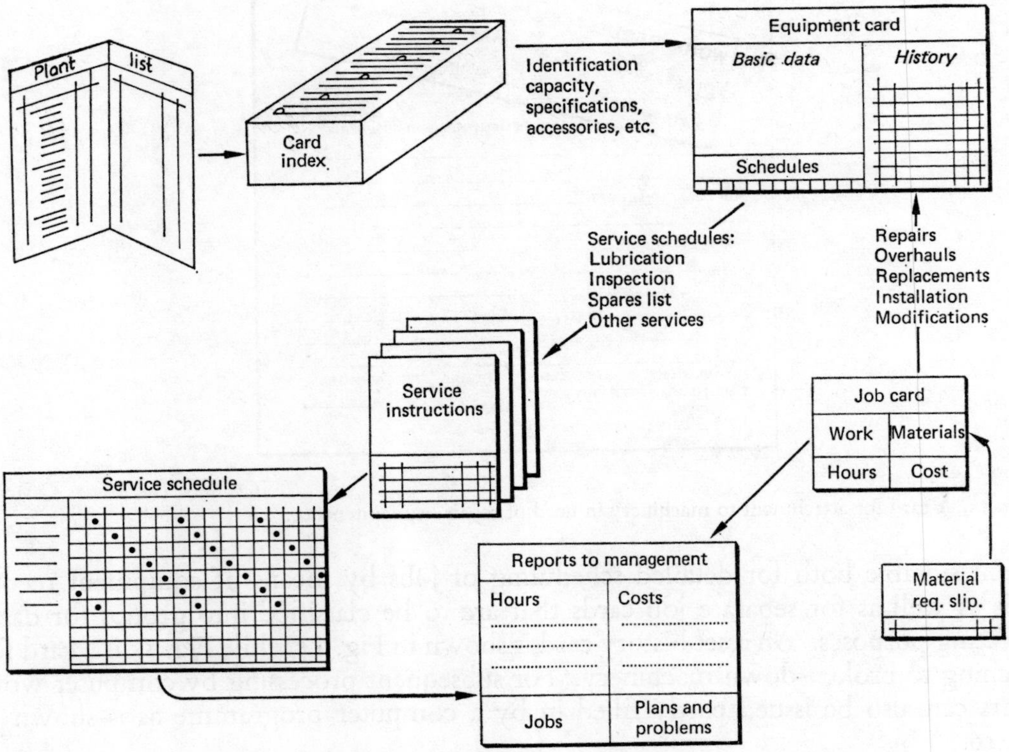

Fig. 17. The basic components and flow of paperwork in the maintenance department.

situation really is? Here lies the justification of paperwork: to instigate action, to record data *and* to give the complete picture. The recording procedure should be supplemented with charting which helps in interpretation.

In this chapter descriptions of the most commonly needed forms will be presented. The over-all cost in manpower required to provide a satisfactory system is hard to calculate, but it should not exceed 6 per cent in small maintenance groups, decreasing to 3 to 4 per cent in larger ones. This estimate normally excludes payroll and attendance time-keeping, as well as the costing of maintenance jobs.

Recording should be regarded as the means that allow interpretation of results. It would be futile to spend time and money without trying to get some feed-back on the results achieved. These results in turn will enable management to evaluate the benefits acquired in the form of quantified results in terms of hours, output and costs. Invariably, the advantages to be gained from good recording must be sold to management. It is no use simply stating that if so much money is already spent on maintenance anyhow, "we'd better try to find out what we're doing with it." The aims of a recording system are to enable us to do this.

The basic components of maintenance paperwork appear in Fig. 17. This diagram is usually augmented by secondary cycles, *e.g.* the work estimating procedure, the spares purchasing cycle, the request for overtime, etc. Let us ignore these for a while, and concentrate on the basic steps.

THE EQUIPMENT INVENTORY LIST

The equipment list is the first essential document in the recording sequence. It should present the collection of plant data in elementary form. However, in drawing it up we immediately realise that it must be done in an orderly manner otherwise our task becomes impossible. It is shown in a simple form in Fig. 18.

As a first step we must decide how to group the equipment. Obviously we cannot list an elevator, a pump, a drill press and a fork-lift in the same sequence. It is first broken up into buildings, mechanical, electrical, safety or office equipment. Let the mechanical equipment serve as an illustration. This group can then be subdivided:

by *location*, *e.g.* Department A, Department B, etc.;
by *type*, *e.g.* presses, grinders, etc.;
by *manufacturing process*, *e.g.* washing, cutting, packing, etc.;
by *function*, *e.g.* production machines, auxiliary equipment, transportation, lifting devices, etc.

or perhaps in some other particular way. Having chosen one of these it imposes a certain discipline that has to be followed. We have to keep this grouping throughout the plant even though there may arise problems of borderline cases that can be included in more than one group. Some guidelines must then be laid down.

There is a surprisingly high proportion of otherwise well-run companies that do not have a plant list. When it does exist it is often out of date. A good plant list provides a bird's-eye view either by group or by department and it will assist us to answer such questions as the following:

How much maintenance work is expected in Department A?
How many electric motors are there in the packaging department?
Which of the machines need replacing?
Which of the presses needs most attention?
What is the insurance value of our plant?

Plant no.	Description	Serial no.	Location	Year installed	New or s/hand	Condition: perfect / good / fair / poor / to be scrapped

Fig. 18. A plant and equipment list representing data in elementary form.

Two examples can be cited here to show the usefulness of a plant list.

Example I. The author was requested by a company to introduce a preventive maintenance system in its production workshops. After some searching the available list was unearthed and it showed the situation as it was five years previously. Replacements and movements since that time were not shown. Thereupon a new list was prepared showing, among other things, the age and condition of the plant.

It was found that 20 per cent of the machine tools were beyond repair and 35 per cent in a poor condition.

Since the company was engaged in precision work these discoveries alone prompted loud protests. Only an actual count and assessment could convince management that these figures were correct. In order to draw up a plan for regular services, management had to state their policy regarding these units. Were they to be replaced or did management expect them to be brought to a satisfactory operating level by the maintenance crew? A policy decision regarding these units was taken and planning of the maintenance system then proceeded.

Example II. In a certain plant two production departments were forever at loggerheads over alleged preferential treatment by the electrical maintenance workers. A listing of the equipment in all departments showed that of all the horse-power connected to the network one of the disputing departments had 35 per cent, namely the lion's share, compared with the 18 per cent in the other department. When this was discovered, it was decided to delegate a full-time electrician to that department, since it would be justified by the existing workload there. Another outcome was the realisation that the production supervisor in the department that was using 18 per cent of the connected horse-power, tended to overstate his case.

The plant list is not made redundant by the use of equipment record cards; in fact, the two complement each other. Besides, the list serves purposes that a card index cannot.

A well-prepared plant list will enable the development of a numbering code. Whether a decimal system or mnemonic one is preferred is not of prime importance. Whichever is chosen it should allow unambiguous identification of the group, the location *and* a serial number. Decimal numbering codes tend to produce five- or six-digit numbers, whereas P15 or D21 can easily identify machines. A suffix of the letters a, b, c, etc. can serve areas or departments.

It is a good idea to prepare a list in such a form as to allow later corrections, deletions and additions. One way to do this is to have the original on tracing paper and to make blueprints of it. Changes can then easily be introduced.

A plant list will give information on one or more of the following:

- a yearly equipment review for budgeting of the service and replacements;
- a visual presentation of the size of machine groups;
- an annual follow-up of plant distribution;
- the designation of vital plant units in each department;
- the establishment of policies regarding service procedures for different groups of equipment;
- the establishment of a centrally supervisable numbering code.

All such decisions can be guided by the list.

The plant list should be complemented by plant layout diagrams identifying the various items on floor plans by the same code number as the one used in the list. The plant layout should show the electrical network, the compressed-air installation and the steam system, if and where applicable. The combination of these details, perhaps by overlaps of tracings, will avoid a great deal of unpleasant surprises, misunderstandings and false moves. Although it can be argued that: "We know this place like the back of our hand," when a new machine is due we go into the shop with a tape to measure whether the available space will be adequate.

In other instances we may try to cast solid foundations on the floor only to find that the draining pipes are directly underneath. Many teams have had trouble when a certain branch of the electrical system gets overloaded. Adequate detailing and proper identification of all relevant points with the maximum load they can carry will prevent many such mishaps. The benefits derived from the fact that information is at our fingertips will justify the initial outlay in getting these plans prepared.

As mentioned earlier, it is advisable to prepare the plant list on tracing paper (A4 BSI size is ideal) so as to allow changes to be made. Blueprinted copies should be circulated to the departments and functions who may use them to some advantage. (With the advent of a vast range of office copiers blueprinting becomes slightly archaic except for the unlimited sizes that the old process can handle.) Depending on the way the list is prepared it can also serve as a basis for the assessment of insurance premiums. However, it may be useful, in addition, to the accounting section in establishing cost centres, overhead allocation and yearly depreciation.

The distributed lists should be recalled yearly, and when all changes have been added, new sets of blueprints should be issued. Circulation of a list will provide a link between maintenance and other departments and create a better understanding between them.

EQUIPMENT RECORD CARD

Before the advent of individual cards the usual practice was to keep a "plant register" or ledger. In these a page was devoted to each separate unit, headed by the name and type of the machine and its location. The number of the page on which a machine happened to appear provided the identification number for that particular machine. The space below the title was then devoted to recording the repairs, listing the spares required and any other problems relating to that unit. The register used to be inspected by the responsible manager from time to time and often every page was initialled by him.

Apart from this last-mentioned use of the "register," the practice did not have much to recommend it. In fact it often became a nuisance, for when a page was filled to capacity the continuation had to appear on a page at the end of the book. Similarly, if in the first instance the machines were listed in certain groups, new additions had to appear on one of the free pages at the end of the book as well. To overcome this, it was common practice to leave a few blank pages after every group, but how many pages to leave was anybody's guess. Next, slips of papers used to be inserted between the pages for an assorted number of reasons: reminders, order slips for spares, memos and requests. The register soon turned into a greasy and swollen dog-eared monster. When eventually a new register became inevitable a lot of transcription took place and from that moment there was the added problem of remembering at what point the second volume took over. Soon the company had two greasy and swollen dog-eared monsters on its hands.

All this is history and we may now be approaching the stage where computers can easily classify, retrieve and present all requested details: visible record computers (VRCs) appear especially promising. Yet the feed-in of information still requires some kind of recording, whether a punch-card or tape, a magnetic-stripe card or a character read-out card. Recording, in other words, is not yet obsolete.

However, computerisation is still not widely used, in fact, we can hardly expect it to become so. Small- to medium-size companies will always rely on manual methods of recording to a certain extent. As the scale of operations in a company grows it may become practicable to have print-outs of service programmes by computers on pre-printed forms. Work done can then be tabulated and summarised periodically by the same computer. (More aspects of electronic data processing (EDP) are discussed under "Scheduling of irregular jobs" in Chapter 6 and "Stock control" in Chapter 7.)

Returning now to the equipment record card (Fig 19), we have to discuss the many ways in which it can be designed, the purposes it should serve and the ways it can be used.

The basic premise is that every item of plant that has an individual life during which it is to be serviced should have a separate card, sheet or file. Thus, if we have a furnace with a number of injector burners, the furnace requires a card and the burners would have individual cards or at least a collective card for all burners. Similarly a radial drill

EQUIPMENT RECORD CARD

Name of unit	*Serial no.*	*Location*	*Plant no.*
Manufactured by........ .... Local agent........	Year manufactured........ Date installed........ Date disposed of........ Reason........ Scrap value obtained........	Purchase order no......... .... Invoice reference........	
Technical data and description: ☐ Speeds and feeds Capacities kW and hp rating Water, steam and air		Weight........ Dimensions ..../......../........ Total floor space........ Foundations........	

Accessories: Motors, A, V, W, phases, rev/min, hp, type
Pumps, cfm, rev/min
Controls and instrumentation

Financial figures	*Service schedules*
Purchase price .... Cost of accessories .... Installation cost .... Total investment ________ Life expectancy........ Yearly depreciation........%........ Acct. no.........Cost centre no.........	Lubrication instruction sheet no......... Inspection instruction sheet no......... Spare parts list no......... Overhaul instructions sheet no......... Other schedules........ ....

Fig. 19. An equipment record card (schematic) giving technical data, cost and servicing schedules. The reverse would normally contain the history of repairs.

requires a card and the electric motors installed on it require separate cards since they have their individual service requirements. Occasionally they may have to be removed to a different location and replaced by other motors.

Both the frequency and the type of services for these different items necessitate this division. A poor second to this solution would be a card for the main machine listing the attached units and allowing separate spaces for their histories. Another argument in favour of separate cards is the fact that since burners and motors have a shorter life expectancy than the main unit, upon their withdrawal from service their card would also have to be withdrawn.

The form that a card can take is a matter of preference and expediency. A simple card file (in a box) will often do. The visible card index where cards are staggered so that the bottom line is exposed ("Kardex" by Remington) is one variation.

Next, a file or folder may be used for each plant item. This again can take one of several forms, such as those used in suspended or lateral filing systems. The advantage here is that the file allows collections of all the paperwork related to that unit in a single place. This is useful in the case of conglomerates or installations. It allows other material, such as the manuals, to be kept together with the rest of the data.

Another approach is the "Keysort" card with perforations according to a code along its periphery (*see* Fig. 13). However, this application is more suitable for coding of information so as to be easily retrievable in different groupings.

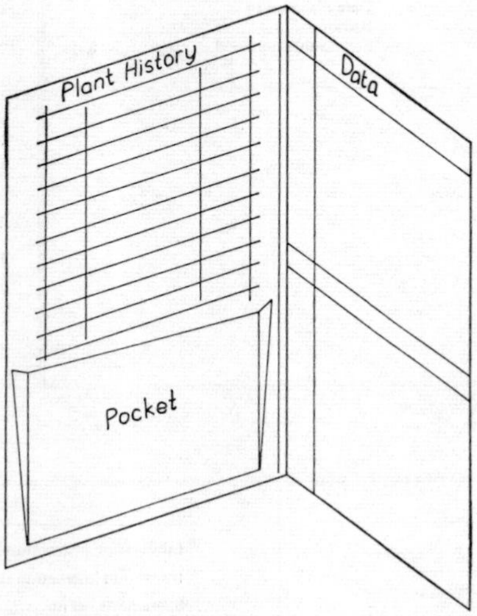

Fig. 20. An equipment record file of the folding type.

There are other ingenious ways of preparing an equipment record card. To accommodate a good deal of writing the card may be A4 in size, folded in the middle and with a pocket attached on the inside (*see* Fig. 20). The pocket can hold purchase slips, job cards or other related notes. A large oil company uses such cards for the maintenance of petrol filling stations. It must be borne in mind that the equipment installed in a station is quite extensive: there are petrol pumps and meters, air compressors, oil pumps, lifting jacks and other units that make up the mechanical side of a petrol station, and then there is the electrical installation. It is thus quite useful to provide all the concentrated information on the whole range of services entered on one folded card.

In this instance it would also be feasible to have a set of cards of different colours inserted into the pocket to serve as record cards for the various groups of units such as pumps, lighting fixtures, buildings and facilities, etc.

Another form of record card is the log-book which is essential for all mobile units, whether passenger vehicles, trucks, boats, cranes, earth-moving equipment or aircraft.

Other equipment such as freezers or accounting machines which may be moved from location to location to serve different masters may also use log-books. Since log-books are fairly common in our motoring age there is no need to describe them here.

There are many good reasons for using a separate equipment record card for each item of plant. Instead of listing the advantages arising out of their use, it should be noted that they serve the following purposes:

(*a*) identification, name, type, number and follow-up of plant location;
(*b*) origin of make, supplier, price and life expectancy record;
(*c*) concise listing of technical data;
(*d*) service instructions reference;
(*e*) maintenance history;
(*f*) activating periodic servicing;
(*g*) initiating the re-ordering of a replacement.

The last two uses are optional and discussion about them appears in Chapters 6 and 7. Experience has shown that in smaller companies it is preferable to keep service instructions apart from the main card on separate forms, although in practice these may remain recorded on the cards. In the author's opinion, the procedures involving coloured "riders" to indicate when some action is due is not as practical as it seems. They are rather in the nature of gimmicks, more time-consuming to operate and tedious to keep up to date than a spread-out schedule sheet. Such systems only become economical in enterprises with about 1,000 plant units or more, where a full-time schedule clerk is assigned to the task. A selection of visible scheduling scales appears in Fig. 47, page 114.

Samples of equipment cards are best used as guides for developing our own forms. In certain countries, mainly in Europe, standard equipment cards have been developed by research centres for typical machinery such as machine tools, electric motors, pumps and elevators. Ready-made forms (*see* Bibliography) come close to satisfying a wide variety of needs, and by being professionally produced and reasonable in cost they save companies a lot of time and money spent on costly experiments.

Since the aim of the present work is to recommend a practical approach to maintenance rather than discuss the principles of form design, let us now deal with the various details and the way they fit into the system. The form appearing in Fig. 19 is used as a schematic example and may be helpful as a reference during our discussion on points (*a*)–(*g*) listed above.

Identification presents no problems except for the plant code number which should coincide with the one on the equipment list. In some instances units may be moved round the plant, and if we insist on keeping track of a certain motor ample space for marking the present location—and subsequent corrections to it—should be provided. Of course, an established procedure for notifying the records clerk of the movement must be observed.

There are many reasons for recording the suppliers of a unit, their agents and other details of the purchase. Without this information correspondence files would have to be frequently consulted and these would in all probability be divided between the technical, accounting and commercial staffs of the company. Details of the original cost including that of delivery, installation and the accessories provided may be kept in the assets files of the accounting department, and are often difficult to obtain when required. Life expectancy indicates the depreciation rate, but it is also a guide for maintenance in adjusting its services and following its overhaul and replacement plans. The main problem is in the recording of the relevant technical data. We invariably refer to the card for physical dimensions, weight and horse-power rating. We may also require belt specifications and details about switches, relays and other accessories. But when it comes to production capacities such data are often not required. Otherwise we would have to transcribe most of the manual—a task that is sometimes needlessly done. To be of help in this respect the following advice is offered: a Xerox or Thermofax copy should be made of the original manual or at least of the relevant pages and kept close at hand. (In fact, additional copies should be made of those service instructions and charts which best serve the PM (Preventive Maintenance) planning section.) As for the original manual, it is best kept away from greasy fingers, in a well-indexed reference library.

We now come to our fourth point which refers to service instructions. Contrary to what has for long been a widespread practice, it is suggested that the equipment record should only carry *a reference number* to the instructions relating to lubrication, inspection and other routines. These instructions should be coded and numbered in a certain sequence and kept in groups, in separate binders (*see* Fig. 23, p. 75). The reference number then allows easy retrieval. Similarly, reference should be made to the spares list advocated by the manufacturer, or to an alternative one developed by the firm which will be kept by the spares stores. This proposal may be in contradiction to some well-entrenched ideas, but enough arguments can be put forward in its favour.

Service instructions can be very detailed and lengthy and refer to at least three different kinds of activities: lubrication, inspection and replacement of parts. The periodicity for these services ranges from hours and days to months and years. To attempt recording all this on the equipment card is to overburden it. It is claimed that efficient coding in figures, symbols and colours can overcome this problem but service instructions are becoming more detailed every year and the cards tend to become a cryptographer's delight.

Now let us assume that all the instructions have nevertheless been duly marked in code on one of an assorted variety of time-scales (*see* Fig. 45) and that we use an old-fashioned card index. Having extracted a card to consult it and transcribe the instructions, the card is replaced and all the visual effects are again out of sight. The only way to locate a card easily is to use a marker or "rider" along the edge of the card. When, in turn, the "rider" plays its role and attracts the clerk's attention the service due has to be interpreted from the code and copied out in longhand since the coloured shorthand coding is meaningless to anybody not involved in the upkeep of the card system. This is evidently a tedious procedure.

The *only* recording along a time-scale that can be justified on an equipment record card is a broad indication of frequencies for the various services or a listing of services due. The three-card system is therefore a step in the right direction. It divides the basic information into (*i*) a data card; (*ii*) a service-instructions and scheduling card; and (*iii*) a plant history card. The system is adaptable and it incorporates features that make for speed and efficiency.

We now come to the mainspring of the card, namely the provision of historical data. While other data are more or less of a static nature and used for reference only, the historical record is used for posting information and interpreting developments. To be most effective and least time-consuming the following principles should be observed:

(*a*) Describe the repairs carried out in concise terms.

(*b*) Record the job-card number so that full details may be obtained via that reference if, and when, required. Work-orders can be kept on file for three to five years in serial number sequence.

(*c*) *Do not record* routine services carried out uneventfully—these should appear elsewhere in a more concise and easily surveyable form.

(*d*) Do not attempt to post on this card the total cost for individual jobs—record only the total hours and main spares used. The value of materials and spares can be obtained later from the stores issue slips whose numbers appear on the work order.

Any attempt to impose a requirement to calculate and post the total cost of the job except in terms of hours is utter nonsense for the following reasons:

(*a*) If time estimates have been made in advance and one wants to ascertain variance of actual from estimate, there is no way of knowing whether the estimate was low to start with or whether the repair was done in a wasteful manner.

(*b*) Whatever has been issued is either on the equipment, on the shop floor or in the scrap-heaps. In the first case the issue proves to be justified, in the latter two it should be detected by supervision and subsequently prevented.

(*c*) The prices quoted for materials issued is both inaccurate and, as far as the department's operation is concerned, irrelevant to the repair job. No stock of spares is ever up to date on all materials, especially spares bought years ago, which may have either appreciated or depreciated in value.

(*d*) The number of times that anybody will enquire about the cost of a particular job is fairly low and then quoting a broad figure will usually be sufficient. Any craft supervisor will be able to assess whether it was a £25 or a £200 job. Accurate total costs can only be calculated, in any event, by the accounting section, considering the need to include additional expenses and the different hourly rates for different trades.

(*e*) Maintenance expenditure which corresponds to the total of all work performed on every unit is calculated and kept by the accounting section who should provide accumulated totals, and these can then be transcribed periodically on to the card.

(*f*) A column is sometimes provided to record the reasons for the various jobs on the equipment card, sometimes in code. However, it may be just as easy to extract the information from the work orders when required. Recording and analysis of breakdowns is discussed under "Breakdown analysis" in Chapter 8.

(*g*) To separately record the hours spent on a job by the various crafts might sometimes be justified. This should not, however, be part of the regular procedure. When large jobs are regularly undertaken, a posting sheet for that purpose should be used (*see* Fig. 31, p. 88).

By keeping to these guidelines the history of a machine can easily be maintained without requiring too much clerical effort.

THE JOB CARD

The purpose of a job card (also known as a "work order," a "work ticket" or a "repair order") is to implement the work to be done and to record the relevant details as they occur. Figure 21 shows how this is achieved. The work order constitutes the link between the equipment in need of service and the organisation as a whole and, in particular, the departments involved in providing the service, *i.e.* maintenance, stores and accounting.

The job card appears in many shapes and variations ranging from simple 3 inch by 5 inch memos, to elaborate forms with tear-off tags or fold-outs. Some may be simple to fill in, some may be complicated forms with much of the details pre-printed, still others may be on punch-cards. None of these versions should depart from the original purpose of linking the phases from the original request to the final interpretation of assembled data.

There are many advantages in having written work orders. Companies which argue against them for the sake of expediency and economy tend to forget that on jobs costing between £15 and £50 the additional cost of time spent on filling in the form is negligible. In return the requestor is able to specify the details of the request unambiguously (so it is hoped) and often in less time than oral explanations would take. The date of the request, a time estimate, if done, and the requested completion date allow a certain degree of planning and scheduling and this is conducive to minimising friction which is all too frequent when instructions are given verbally. There is also "historical" value in the various dates that would appear on the job card, if properly analysed. Finally the data accumulated on the job cards serve as a basis for controls.

A request written on a form is more easily handled than if the information were only in the minds of the people concerned. It can be placed in a certain sequence, transmitted without change and it serves as a reminder of jobs to be done and of details to be recorded. Having passed through all its stages a job card becomes a document that knows and remembers all and with which there can be no dispute. Last but not least is the fact that it provides the only way to activate the repair service effectively and to provide management with information regarding the expenditure for a cost centre or charge account. Prior to a discussion on job-card scheduling, the experience of a company can be mentioned where manpower utilisation rose by 65 per cent as a result of the introduction of an improved work-order procedure. This resulted from the fact that jobs could now be grouped according to trades and allocated in a more orderly manner.

Jobs that do not require a job order to pass through one or more of the phases shown in Fig. 16 may not require a job card at all. For example, if there is a standing order for a certain operation to be done at regular intervals, or in the case of an occasional adjustment of a machine lasting not more than ten minutes, these jobs do *not* need a work order. Jobs lasting up to ten minutes (this practice may differ in various factories) can

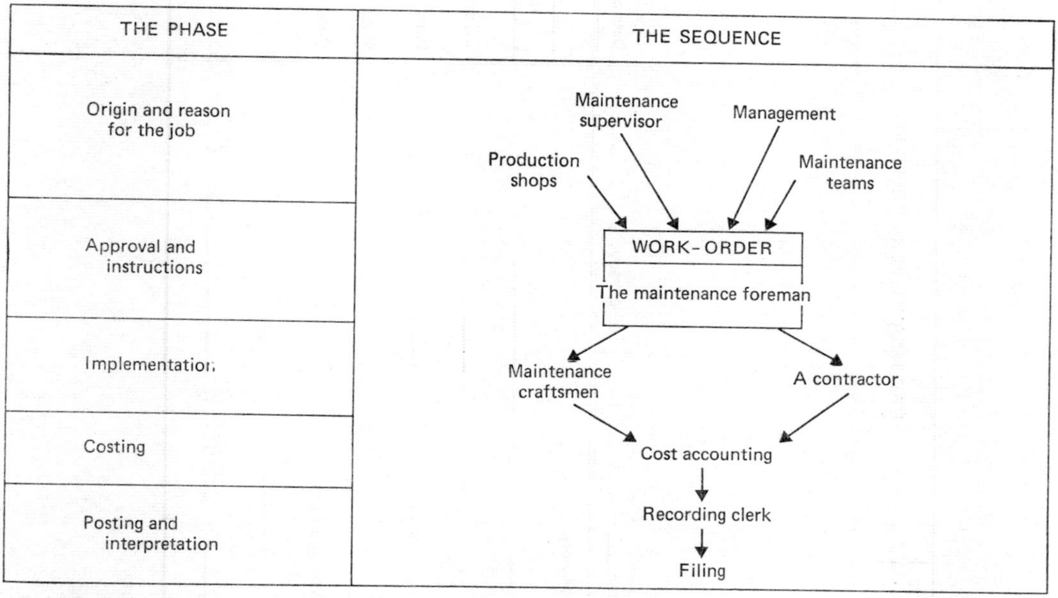

Fig. 21. The work order's progress through a sequence of stages.

be accumulated on a weekly card made expressly for that purpose. Regularly recurring jobs should appear on schedules or pre-printed forms where only a check mark indicating completion is necessary. These will be discussed in a subsequent chapter. It thus becomes clear that job cards are issued for *work occurring irregularly* or for jobs which occur *at long intervals* or, finally, *regular but very costly jobs, e.g.* renew mixer lining after every 4,000 hours of operation. In other words there must be a good reason for putting the request on paper and insisting on its passing through all appropriate stages.

The itinerary that a job card follows and the information for which space is provided affect each other. Let us then follow the progress of a work request that comes to the maintenance supervisor from a production foreman. Let us assume that we dispense with a cost estimate and a budget authorisation. A job card of this type would look approximately as shown in Figs. 22a and 22b.

The information requested on this form will fall into the groups shown in Table V.

Acct. no.	Details	
	Shop no. 1	
	Shop no. 2	
	Assembly shop	
	Power house	
	Maintenance	
	Shipping	
	Transport	
	Emerg/repair	
	Installation	
	After/insp.	
	Development	
	Overhaul	
	Fitter/mech.	
	Electrician	
	Builder	
	Painter/carp.	
	Labourer	

IMMEDIATE		MAINTENANCE WORK ORDER NO	
REGULAR			Plant no.

Date ordered........................ To be ready........................ Date completed........................

Details of work: ....

....

....

....

....

Remarks/suggestions: ....

....

COST SUMMARY

Items	Est.	Act.
Labour hrs.		
Purchases		
Matls. issued		
Contracted, etc.		

Fig. 22a. The maintenance work-order form, front.

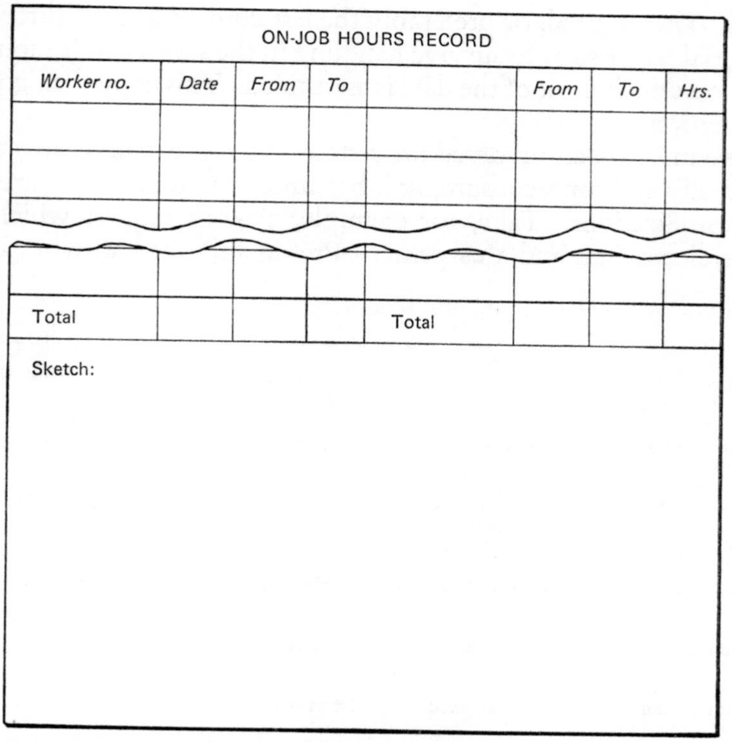

ON-JOB HOURS RECORD

Worker no.	*Date*	*From*	*To*		*From*	*To*	*Hrs.*
Total				Total			

Sketch:

Fig. 22b. Reverse side of the maintenance work-order form.

Should the procedure normally require an estimate in man-hours and costs, space should be provided for that purpose. For jobs above a certain sum authorisation has to be obtained and recorded as well, with date and signature of the person authorising its implementation. In some cases the supervisor has to re-check and diagnose the fault on the spot and issue instructions accordingly. This could easily be accommodated in the space reserved for "Details of work." The hour of stoppage and of start-up is sometimes deemed important; however, this detail is best recorded by the machine operators who come under the responsibility of the production department. There is, after all, no way for a maintenance man to know or to find out the exact time of stoppage. Design of the form should be such that each person involved in the procedure should deal with a defined space on it, possibly framed by heavier lines and appearing in the sequence to be followed. The size should allow the card to be inserted into a breast-pocket and both sides of the form should be used. Space provided for an eventual sketch is usually wasted and it seems preferable to attach a separate sheet for that purpose. Occasionally, though, there are exceptions to this rule.

The number of copies to be made varies from one to five. There is a good case for the requesting person to keep a copy and there is a good case for the maintenance foreman or scheduling clerk to get a copy. Then there is the working copy on which instructions, hours, dates and materials issued are recorded. Additional copies are not recommended. The original, or preferably the last copy of a set of three, should be on a more substantial paper so as to survive handling in the shops and it should serve as the working copy on which most of the data is assembled. This copy then goes to accounting for computation.

The more we insist on an involved procedure, the more details have to be recorded and the most costly and time-consuming it becomes. Naturally there are factors which give rise to complications. Take, for example, a work request which involves the mechanics, the electricians and a carpenter. Presumably they all work on the basis of

Table V. Information required for a work-order form.

Identification	*Action required*	*Action taken*	*Computation*
Department or shop	Actual problem	Work done	Total cost of man-hours
Date requested	Apparent reason for defect	Work identification	Cost of stores issue
Date to be ready	Work instructions	Number of hours	Cost of purchases
Plant no.	Class of job	Dates worked	Total hours
Plant type and name	Trade involved	Stores issued no.	
Urgency	Authorising signature	Local purchase no.	
Form serial no.		Date completed	
Charge and account no.		Acceptance check	

job cards. How are we to get enough copies to go round? An obvious solution, of course, is by using a copying machine. Temptation is also great to exceed the basic number of copies needed or recording more details than are strictly necessary. One often finds that the form provides spaces for dates indicating its passage from one craft to the other and from one stage to the next. The notifying of everybody concerned is another pitfall. The number of persons to be notified should be carefully limited, otherwise the time of too many people is wasted.

To facilitate use of the form or the subsequent handling of the data a number of "tricks" can be applied. For instance, job cards can be printed on paper of various colours with each production department or factory area assigned a different colour. This makes it easier to keep track of an order and facilitates the collection and classification of data for reporting purposes. Bold lettering may be used to designate first, second and third copies and the destination of each.

Another measure that helps both the writing of the order and its subsequent interpretation is the pre-printing of frequently recurring data as in Figs. 22a and 22b, in which the appropriate boxes have to be crossed at the time of issue. Similarly, the pre-printing of degrees of urgency (immediate, urgent, normal), the reasons leading to the breakdown (neglect, improper use, accident, inadequate lubrication, normal wear and tear) and the classification of orders (repair, modification, investment work or planned service) are particularly useful. By stacking completed forms at the end of a period into the respective groups, we can summarise monthly frequencies, total costs and man-hour expenditures for each group. This is an essential part of the control stage. Without the pre-printing each completed job card would have to be fully read and interpreted before grouping. Pre-printing can thus save considerable clerical time.

A similar practice is that of using code numbers for frequent repairs where many identical units are in operation, *e.g.* replace fuse, adjust belt tension, etc. These can be grouped into mechanical, electrical and miscellaneous jobs, using a prefix for each group. The code for these standard requests can appear in fine print on the reverse of the form.

A frequent mistake is to try to identify different jobs by a series of code figures, referring to class, type, group, category—and so on. The most important coding is the plant number and the next important is the job card serial number. For regular groups of jobs a thorough system of account numbers should exist.

Furthermore, the request and issue of materials from stores can be made on forms bearing the same *serial number* which are detached from the job card. Thus every work order has a materials issue slip directly related to it which can easily be located for accounting purposes and does not require transcription of identifying data.

A well-designed job card deserves a well-planned procedure and this in turn can only be ensured if all concerned know its uses and comply with instructions. A thorough discussion and demonstration will certainly prove useful. Among other things it has to be explained to potential requestors that a concise and clear description of the trouble is desirable. "At 10.30 the machine broke down!" is not very helpful. Speaking from experience in a country with many immigrants (Israel) and a technologically developing country (South Korea), the teaching of a common terminology has been conducive to much improvement in this respect and to the great benefit of everybody.

Another interesting point which is often debated is the designation of the urgency of a job. Many "old hands" say that given half a chance every requestor will demand that his job be given top priority. This may be a fact or a guess but it does not mean that this will always be so. When this point is brought out in a discussion it soon becomes clear that all jobs cannot be given top priority since there is not enough manpower to deal with them. (A very sensible priority list has been suggested on p. 367 of E. T. Newbrough's *Effective Maintenance Management*, McGraw-Hill, N.Y., 1967.) The point is soon taken when the most vociferous complainant is asked to imagine himself in the role of the maintenance supervisor. It often happens that the person requesting a job has delayed for so long a time that it has become an "emergency" and he then calls upon maintenance to bail him out. (For further discussion of priorities, *see* "How to deal with the workload" in Chapter 6.)

To summarise, the job card must be laid out in such a manner as to cause the requestor to fill in his part in an easy-to-follow sequence, without omitting any important detail. The supervisor receiving the form can then interpret the request, verify the description, make his own remarks and indicate who should do the work. The scheduling clerk (despatcher) will allocate the job, according to capacity, to one of the craftsmen by placing the job card in the worker's slot on the board (*see* Fig. 55). The worker takes the working copy of the card and records on it the starting time and date. He leaves a control copy for the progress board. He keeps filling in the times and dates when he is working on the job on the working card until the job is completed and then returns it to the despatcher. The time taken on the job is then summarised and a short description and the total hours worked are transcribed to the equipment record card. The job card then goes to the accounting office which, after having compiled the total cost and posted it against the appropriate cost centre account, returns it to maintenance. Completed job cards should be filed according to their serial number or requesting shops or by plant number.

In some applications the work order may have a detachable scaled strip bearing the identification number of the job. This strip may be cut to a length corresponding to the estimated time of the job and inserted in a load chart. The length of the strip is additive and it will indicate the total advance load for a craft group or craftsman.

It is unavoidable that variations in this procedure should occur and questions will then arise, such as the following: "How should we proceed when two or more crafts have to work on the same job?" "How do we co-ordinate their work and how do we notify them?" The answers to these questions must be worked out in detail and one must be certain to allow for such eventualities.

The procedure for cases of extreme urgency needs to be laid down and explained. Naturally when emergencies happen we must first attend to them and fill out forms afterwards.

To avoid charges of rigidity in procedure the maintenance staff should be instructed not to refuse small jobs on the pretext that a request must first be filed. Whenever this happens and assistance is refused the person requesting it will feel antagonistic. This is detrimental to good working relations and should be avoided.

The next section deals with work orders that go through a different procedure from those discussed here, *i.e.* they are repetitive and are part of servicing schedules that do not need separate job cards. Basically, work orders initiate work to be done. Within this definition we can include checklists that are tied to certain dates and are repeated without change, and also scheduling charts indicating the dates prescribed for a service. In the cases of routine service these *also* serve as work orders. However, for the sake of clarity they are dealt with in the section relating to service instructions.

So far the problem of time-based incentives and the procedure for estimating job times have been left out on purpose so as not to cloud the issue. The form itself can easily be adapted for that purpose and it does not alter the job-card procedure in its essential details.

STANDARD PROCEDURES AND SERVICE INSTRUCTIONS

The preparation and presentation of service instructions is an art in itself. One has to tread carefully on the dividing lines between the service recommended by the manufacturer, the demands of actual working conditions, the company's replacement policy and common sense. To be on the safe side some companies go by the principle: "First maximise, then minimise!" *i.e.* do more than is absolutely necessary to start with and when you have gained experience reduce the requirements.

One thing remains certain: if expenses are not to soar and costly errors are to be avoided the best man must be chosen for this job.

Service instructions begin taking shape when the machine is being ordered. We must ask ourselves: "Can we provide the service that this machine will need?" Next we check on the details of installation. Are we creating problems from the start by installing the machine in a certain way? Are those problems avoidable? If indeed we can see problems coming, are we taking steps to solve them? In fact, anticipating problems is the best way of minimising them. Hence the well-founded argument that maintenance people should participate in all decisions regarding the purchase of machinery.

Another problem is that of obsolescence. Certain machines are bought and installed with a definite time limit in mind. Certain pieces of equipment have a foreseeable life; for others we may choose to establish a useful life-span. The amount of maintenance service to be provided must take this estimate into account or otherwise over-maintenance may result.

It is suggested that every enterprise should follow certain procedures for ordering, receiving and installing equipment. This is the meeting-ground between "plant engineering" in its proper sense and maintenance. Whatever mistakes are made by the former the latter has to live with them and overcome them. A great deal of unnecessary work could be avoided if, as suggested, sound principles of collaboration were observed.

Standard procedures

Standard procedures differ from service instructions in that they are common to all units and they relate to the time before the equipment goes into operation up to the time when its useful life is considered over. Activities which can be performed by standard procedures are shown in Table VI.

These procedures are given as an example for a metalworking company using mostly machine tools. In a chemical plant, instructions would differ in some details. However, in every case the wording should be concise with the emphasis on verbs denoting action. Once instructions have been laid down in this form, they will prove to be very helpful.

Table VI. Examples of standard procedures.

THE PHASE	*WHAT IS TO BE DONE*
Ordering and approving purchase	Check machine specification
	Compare utilisation capacity to demand
	Check for novel operating features
	Check for h.p. rating, floor space requirements
	Ascertain limitations in use
	Compare with existing equipment
	Make mental assessment of maintenance needs
	Request essential spares
Receipt and acceptance	Check receipt of all items
	Inspect for eventual damage
	Read owner's manual
	Check for protective measures, lubricants, coating, anti-vibration packing
	Check wrapping and packaging for components
	Preserve packing-list and other documents
	Open separate file, if one of a kind
	Prepare minimum spares list
	Assign plant no. and fill in equipment record card
Installation and test run ("commissioning")	Check proposed location for space, electrical connections, air attachment, drainage run-off, etc.
	Issue work order for the installation
	Prepare area and move equipment to location
	Lubricate as per instructions
	Inspect connections and test-run
	Manual in hand, verify functions of all operating components
	Test-run until proper sequence is established
	Instruct operators in all details
	Draw up operating and emergency instructions
	Apply safety markings
	Complete and sign job card
Transfer of equipment	Request written authorisation
	Inspect new location
	Plan removal procedure and timing
	Transfer and install
	Complete and return job card
	Mark new location on equipment card
Disposal and salvage	Obtain written authorisation and instructions
	Remove equipment and protect exposed connections
	Deposit and store, well protected
	Request instructions to dismantle, salvage, overhaul or scrap
	Obtain appraisal for residual value
	Complete equipment record card
	Complete and sign job card
	Notify stores for spares disposal

Service instructions

Service instructions relate to regular, repetitive operations that should be carried out during the lifetime of the equipment. These include:

(*a*) lubricating instructions;
(*b*) inspection routines;
(*c*) periodic adjustments;
(*d*) "preventive replacement" of components;
(*e*) cleaning and protective measures;

(*f*) instructions for overhauls;
(*g*) spare parts lists.

Figure 23 shows the recommended method for handling service instructions. Different groups of instructions should be on separate sheets and each group filed in a separate binder. These can then be handed to persons responsible for the various services. The spares list binder, for instance, should be made available to the spares

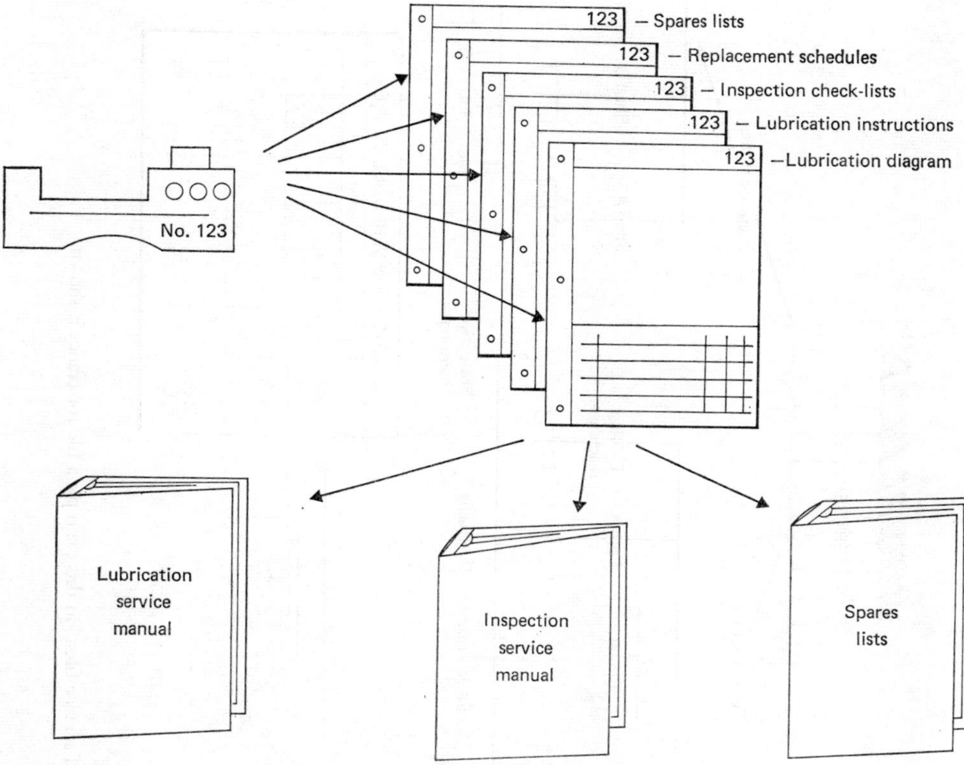

Fig. 23. Recommended method for handling service instructions. Different groups of instructions have separate sheets, and each group is filed in a separate binder, since they will be carried out by different teams.

stores and a copy kept with maintenance. Since they are only "recommended" lists they have only reference value. However, a final and approved list must also be to hand for stock-control purposes.

Looking in greater detail at lubrication services the following procedure emerges for each machine: usually there is a diagram of the machine indicating lubrication points and a list explaining how each point should be serviced. There may also be a sheet where the points have been grouped according to required periods of attention. The complete set of sheets referring to all the equipment constitutes the lubrication binder.

Thus the information contained in this binder provides the instructions for drawing

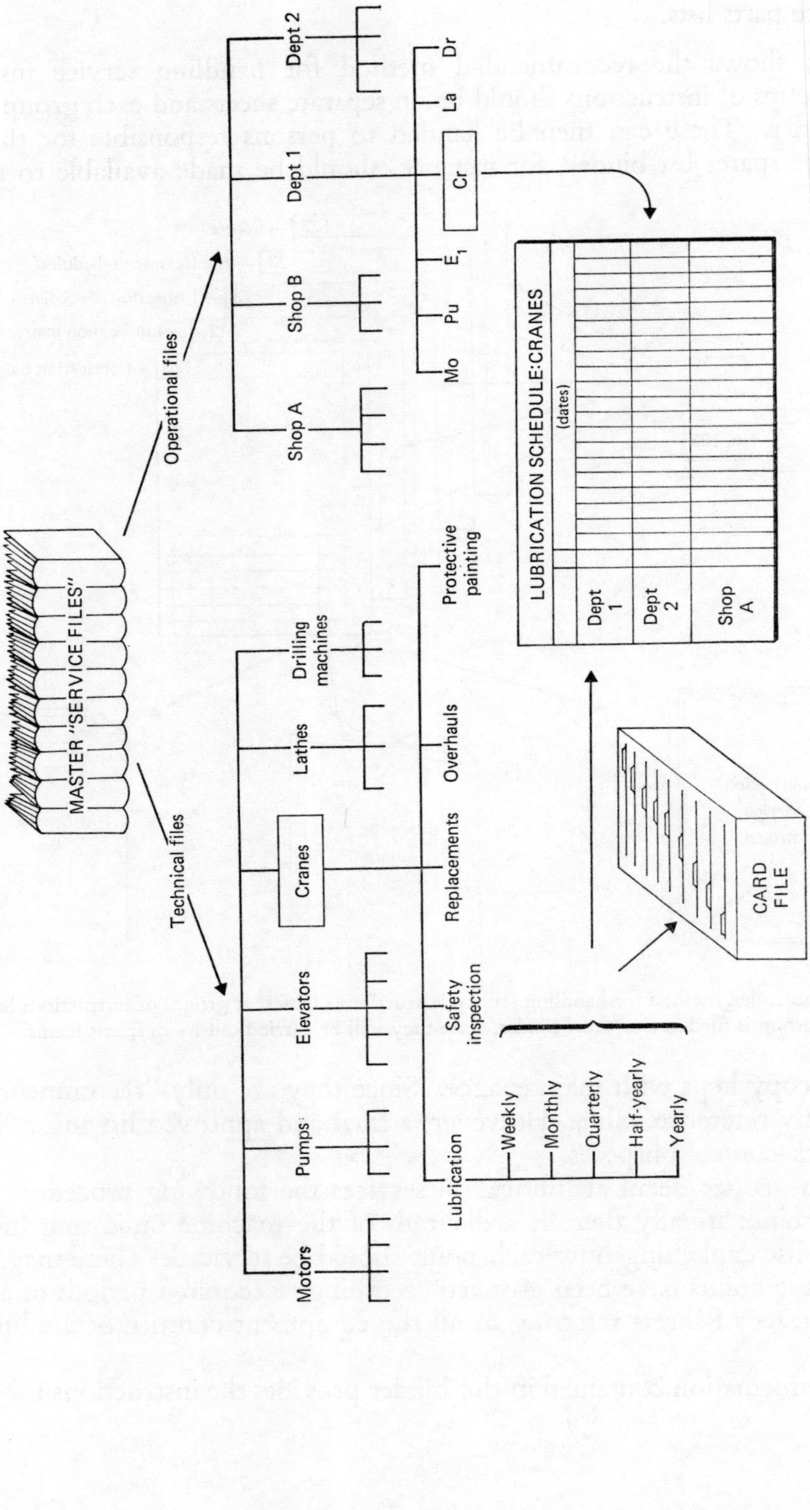

Fig. 24. Schematic arrangement of service files. In this example a file for cranes is shown.

up a lubrication service schedule in chart form. This task would have been immeasurably harder to do from manuals and from the equipment card and in the latter instance would have caused prolonged interference with the card system. A separate binder holding duplicates of all instructions arranged *by sequence of plant number* is optional. The master files are arranged and grouped as shown in Fig. 24.

For each machine the total of these instructions represents the workload on the maintenance force resulting for that particular unit. Since these services have to be provided regularly it would be best to map them out on a time-scale covering at least one year. It should not be too difficult to assess the yearly workload in man-hours which is required by a unit. This schedule should contain space for recording the estimated service times. Worked-out examples, Figs. 32 and 38, appear in Chapter 6.

After several years of operation the machine will be overhauled, and a new cycle is started. The overhaul service has been left out of the annual calculations of time estimates since it will have to be scheduled into the non-regular workload.

These services have been grouped into different categories to emphasise the fact that they may not all be required on one unit and that they may be applied separately. On the other hand, the separate listing of services does not mean that they should not be combined when implemented. This is to be decided on the basis of expediency. One guideline in this respect is the skill and qualification of the service team. A further division may occur because the tools, materials and techniques involved differ, so that separate teams are required. It must also be decided whether production operators may perform part of these services.

Checklist type of service instructions

Service instructions often take the form of checklists, which serve as a guide to the performance of a certain sequence. Their implementation does not require highly qualified personnel, and markings by colour or number code on the equipment should facilitate a correct performance. Most commonly checklists refer to lubrication or inspection, and as such require a check-off space for each point. A descriptive assessment may be required, and for that purpose space should be provided against each point. To compile checklists the machine manual should be consulted and the prescribed service frequencies and details should be adapted to the particular needs and conditions of plant utilisation. Past history of similar units should be taken into consideration.

Each checklist should identify the machine type and name, the periodicity of the service in terms of running hours, tons of output, mileage or calendar periods, the craft group required and the estimated time for the job. A diagram or photograph duly protected by a sheath may be provided to facilitate the work. Upon completion the worker should sign the form and record the date. There are two types of checklist: one that allows recording and has to be reissued every time, and one that requires no recording, is permanent and which must be accompanied by another form to serve as a job card.

Assuming that we desire a checklist on which the service is to be recorded, we distinguish between the following alternatives.

A. A form for *one* unit and *one* occasion. Completed forms are then filed.
B. A form for *one* unit, for *repeated use* during a predetermined period. This form is reissued every time the service is due. The completed form is filed at the end of the period.
C. One form for *several* identical units, for *one* occasion. The form is filed upon completion of all units.
D. One form for *several units* for *repeated use* during a certain period. Forms are filed for reference at the end of the period.

Example of blanks for types A, B, C and D appear in Fig. 25. The difference is in the provision of additional columns, by which a form can be adapted for reissue at regular intervals. In a similar way it can also be made to serve more than one machine. (The letters A, B, C and D will be used to designate types of forms in subsequent illustrations of this book.)

Bearing in mind the clerical workload, Type A is the least efficient. A new form has to be issued for each machine every time the service is due. Referring to Fig. 26 we can easily follow the sequence of events. Assuming a weekly, a monthly and a semi-annual service, there will be sixty-six (52 + 12 + 2) forms completed in a year for one machine alone. If ten units have to be serviced the total will be 660 forms.

It can be rightly argued that the monthly service should appear as an addition on twelve of the weekly service forms and the semi-annual service would be added on two of these. In that case we would have 520 Type A forms per year for ten machines (forty weekly services, ten weekly + monthly services, two semi-annual services, all multiplied by ten) (*see* Fig. 26, Var. I).

By using a Type B form for the ten units (providing that the same service is performed), covering six-monthly periods on each form, we will only have forty forms by the end of the year—two forms per year for the weekly service for each of the ten machines equals twenty and two forms per year for the combined monthly and semi-annual service equals twenty (*see* Fig. 26, Var. III).

If we use Type C forms each for a group of ten units we get sixty-six forms per year. Using Type D forms and combining the weekly and monthly services, only thirteen forms will be needed in one year.

In combining the services on one sheet we create a memory problem as illustrated in the variants of Type A, namely, we have to remember which form is to be issued every week since they are not identical. Therefore a combination of services as in the Type B form is preferable where the same forms are repeated throughout the year. Examples of the applications of these forms appear in Chapter 6.

It is interesting to calculate the number of forms that we will use annually when services of different frequencies are combined on to one sheet. When one type of service only appears on one sheet the calculation is easy. For example: it is self-evident that twelve sheets would be required to implement monthly services or six sheets per annum

Type A

SEMI-ANNUAL INSPECTION CHECKLIST | Code no.

EQUIPMENT .... | Location

No.	Instructions	Craft	Check	Correction	Remarks

Type B

SEMI-ANNUAL INSPECTION CHECKLIST | Code no.

EQUIPMENT .... | Location

No.	Instructions	Craft	January 70		July 71	
			Ch.	Correction	Ch.	Correction

Type C

SEMI-ANNUAL INSPECTION CHECKLIST | Code no.

EQUIPMENT .... | Location

No.	Instructions	Craft	Boiler no. 1		Boiler no. 2	
			Ch.	Correction	Ch.	Correction

Type D

SEMI-ANNUAL INSPECTION CHECKLIST | Code no.

EQUIPMENT .... | Location

No.	Instructions	Craft	Boiler no. 1		Boiler no. 2	
			Jan.	Jul.	Jan.	Jul.

Fig. 25. Four types of form for service instructions allowing repeated use either for one or for several units.

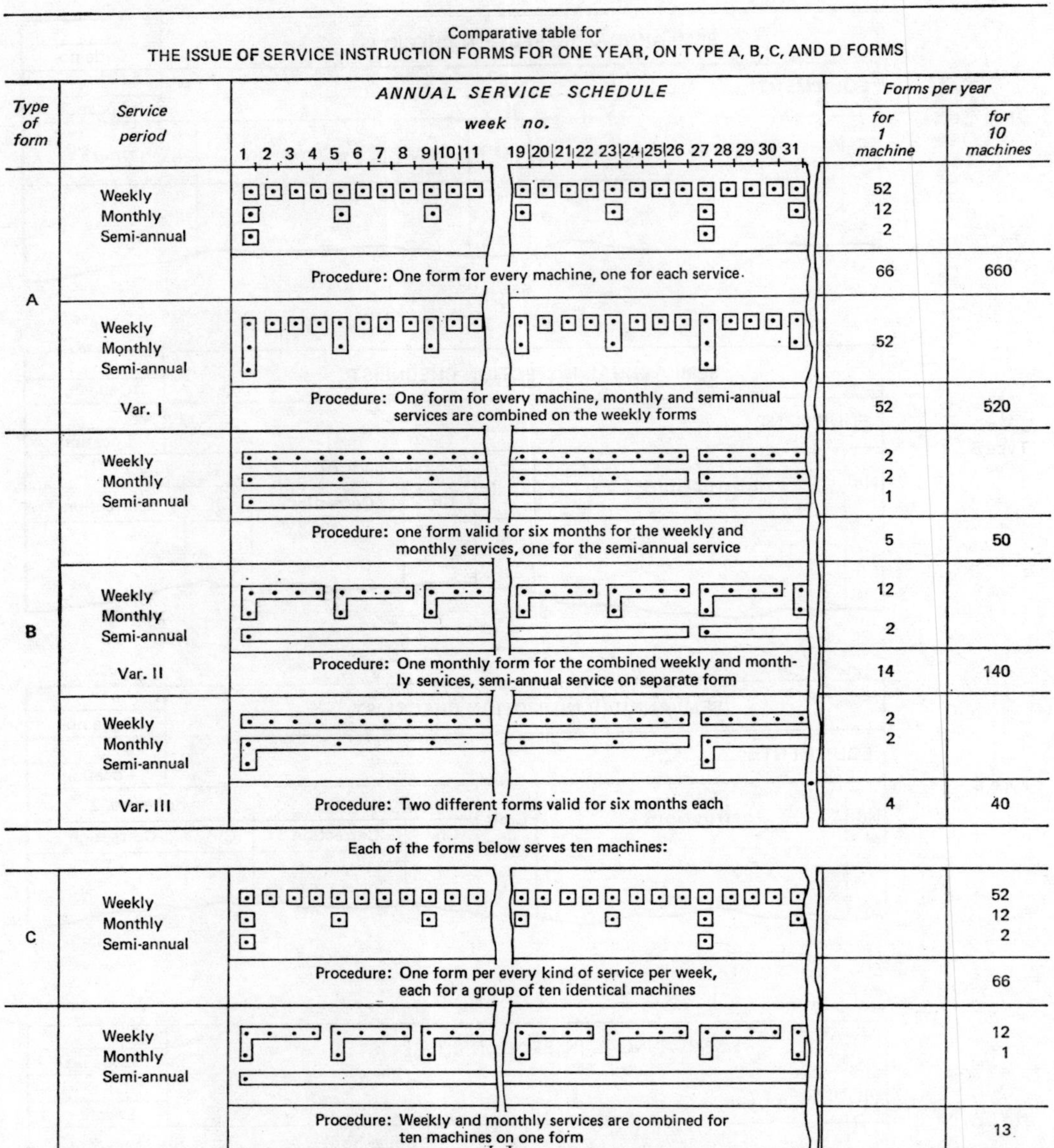
Comparative table for
THE ISSUE OF SERVICE INSTRUCTION FORMS FOR ONE YEAR, ON TYPE A, B, C, AND D FORMS
Type of form
Service period
ANNUAL SERVICE SCHEDULE
week no.
1 2 3 4 5 6 7 8 9 10 11 19 20 21 22 23 24 25 26 27 28 29 30 31
Forms per year
for 1 machine
for 10 machines
A
Weekly 52
Monthly 12
Semi-annual 2
Procedure: One form for every machine, one for each service. 66 660
Weekly
Monthly
Semi-annual
52
Var. I
Procedure: One form for every machine, monthly and semi-annual services are combined on the weekly forms 52 520
Weekly 2
Monthly 2
Semi-annual 1
Procedure: one form valid for six months for the weekly and monthly services, one for the semi-annual service 5 50
B
Weekly 12
Monthly
Semi-annual 2
Var. II
Procedure: One monthly form for the combined weekly and monthly services, semi-annual service on separate form 14 140
Weekly 2
Monthly 2
Semi-annual
Var. III
Procedure: Two different forms valid for six months each 4 40
Each of the forms below serves ten machines:
C
Weekly 52
Monthly 12
Semi-annual 2
Procedure: One form per every kind of service per week, each for a group of ten identical machines 66
D
Weekly 12
Monthly 1
Semi-annual
Procedure: Weekly and monthly services are combined for ten machines on one form 13
Weekly
Monthly
Semi-annual
12
Var. IV
Procedure: Weekly, monthly and semi-annual services are combined into one form, each serving ten machines 12

Fig. 26. Schematic presentation of the use of type A, B, C and D forms when service schedules of different frequencies are combined. Every dot represents one service, every frame one form.

are needed for six bi-monthly schedules. When the same form also contains instructions for monthly or semi-annual services the number of sheets per year is not so easy to calculate. To illustrate this point, let us consider the two following cases:

Case I. For fifty-two weekly services there will be fifty-two forms. However, out of these there will be twenty-six forms for combined weekly and fortnightly services. When there are monthly services as well, only twenty-four weekly forms will be required—sixteen fortnightly and twelve monthly forms. The annual total of forms will be fifty-two since weekly services are of the highest frequency.

Case II. When a machine requires fortnightly, monthly and semi-annual services, there will be fourteen fortnightly, ten monthly and two semi-annual forms, bringing the total to twenty-six forms per year. These figures can be obtained by empirically following Fig. 26 (*see* Table VII).

Table VII. Number of forms required per year when services of different frequencies are combined in one form.

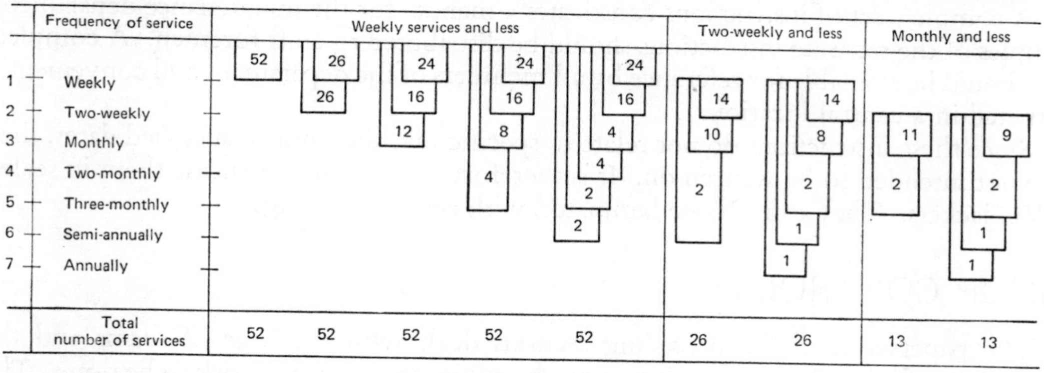
Frequency of service
Weekly services and less
Two-weekly and less
Monthly and less
1 Weekly
2 Two-weekly
3 Monthly
4 Two-monthly
5 Three-monthly
6 Semi-annually
7 Annually
52
26
26
24
16
12
24
16
8
4
24
16
4
4
2
2
14
10
2
14
8
2
1
1
11
2
9
2
1
1
Total number of services
52 52 52 52 52 26 26 13 13

In the final count the number of forms that are issued for the performance of regularly recurring services depends on how the services are combined. The factors that contribute towards an increased number of forms are: (*i*) the number of craft teams; (*ii*) the annual frequency of services; (*iii*) the diversity frequencies; and (*iv*) the number of different machine groups. Needless to say, a proliferation of forms is not advantageous since retrieval becomes more time-consuming and space is taken up by the accumulated paperwork.

Work specifications

Work specifications define the ways and means of doing certain jobs so as to ensure that they are carried out in a uniform manner. It is definitely an advantage to observe certain standard prescriptions in performing certain jobs. These instructions serve as a guide for the tradesman and enable him to comply with prescribed practices, *e.g.* wiring procedures, marking of pipework, levelling of machines and so on.

The various jobs to be specified range from instructions regarding the installation of equipment to notification of the supervisor in case of an accident. Similarly, dismantling of machines, scrapping of parts, salvaging and reconditioning of units, should not be left to the sole initiative of individuals and to decisions made on the spur of the moment. We can thus distinguish between the following sets of standard instructions.

(*a*) Instruction relating to all plant.
(*b*) Instruction relating to groups of machines such as the air-conditioning system, the water plant and boilers.
(*c*) Instruction regarding common procedures, such as wiring, numbering, colour coding, etc.
(*d*) Administrative procedures, such as reporting accidents, requests for shutdowns, recording of repairs, etc.
(*e*) Instruction for start-up and shutdown procedures, operating, cleaning and painting.

A complete set of instructions constitutes a manual for the maintenance department. Copies of the separate instructions should be distributed to craft foremen. A complete set should be available for reference by all members of the department and conveniently situated in a central location.

Since these instructions do not relate to specific machines nor to specified dates, they are not intended to be written on. It is therefore a good idea to encase them in sealed PVC jackets. They can also be laminated with acetate sheeting.

SHOP CONTROLS

The paperwork in the preceding sections dealt with the issue of forms and the recording of jobs. The emphasis was on the equipment and the work to be done. The foreman, however, deals with *people* and their working *time*. To relate the job to be done to the people who implement it, to plan it in advance and to record the details of implementation as the job progresses is known as "shop controls." Work planning and scheduling *precedes* shop controls and management controls *emanate* from shop controls. This concept appears in Fig. 27. Management controls are impossible without the vital link of shop controls. Attempts to exercise control at management level without a firm basis of data from the shop floor will always be futile.

With regard to Fig. 27, however, the following point must be emphasised: it is often mistakenly assumed that the exercise of controls requires additional clerical work and special figures. This is only partly true. The only requirement is that first-hand shop-floor figures should be posted correctly and in such a way as to facilitate computation. In fact, compiling the data after it is recorded is the *only* extra work involved. The interpretation of data and its presentation in meaningful form is discussed in Chapter 10. The clerical work leading up to that stage and the necessary paperwork is discussed below.

Work orders provide "control" over individual jobs. They accompany the job from inception to completion. Service schedules control repetitive cycles. Both of these

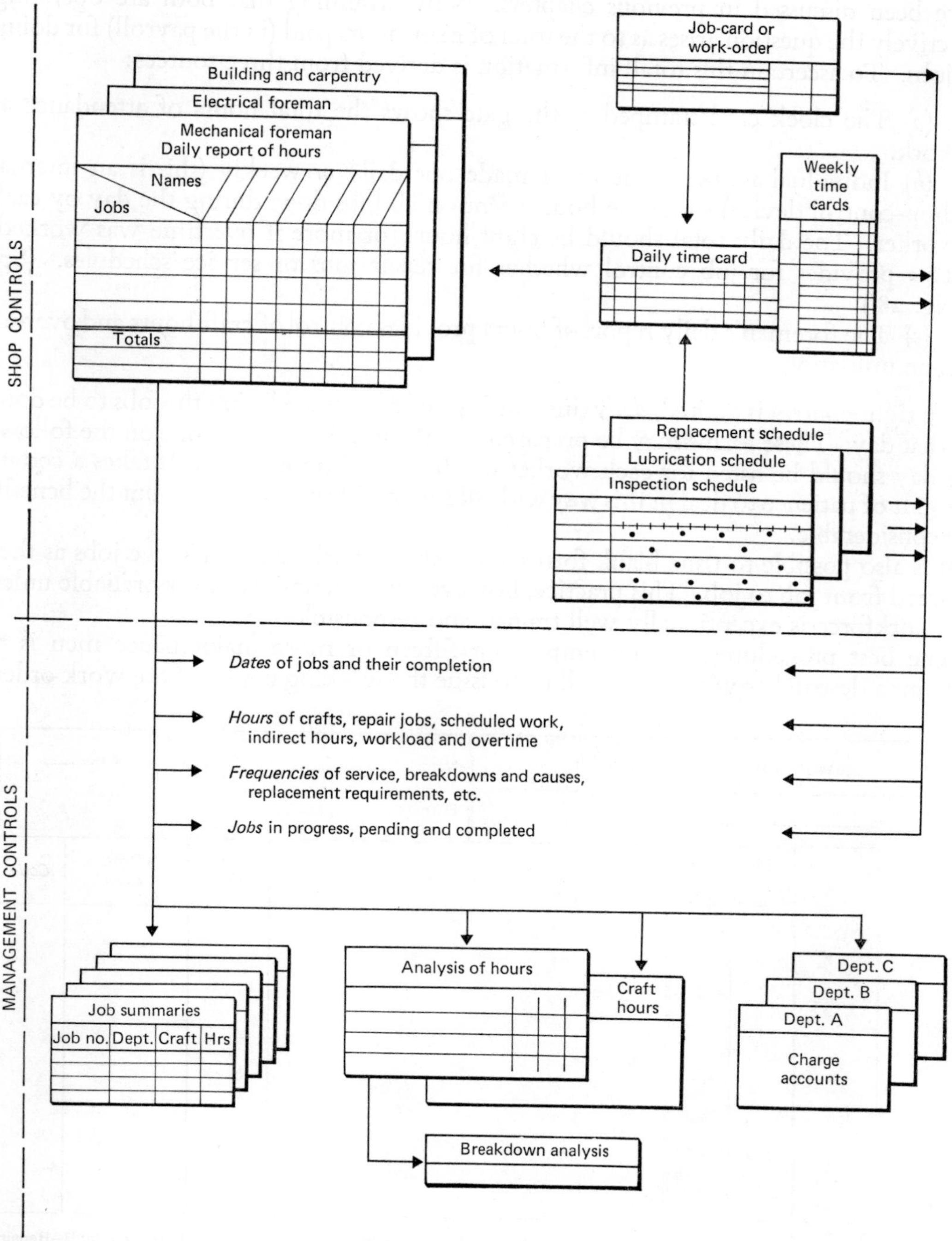

Fig. 27. The flow of data from shop controls to management controls.

have been discussed in previous chapters. Now, assuming that both are operating effectively the question arises as to the total of man-hours paid (in the payroll) for doing all jobs. To ascertain this total, information is derived from three sources:

(*a*) The clock card stamped at the gate shows the total hours of attendance at work.

(*b*) Individual worker time cards made out daily *or* weekly (this is an internal shop-control device) show the hours spent on all jobs done during the day by each worker. The daily total should be eight hours, or more if overtime was worked. This provides for job control whether for repair jobs or service schedules. (See Fig. 28.)

(*c*) The foreman's daily report of hours provides control of craft hours and over-all time utilisation.

If a tight control is desired, *daily* time cards should be issued listing the jobs to be done on that day. These forms may be prepared in advance. Jobs to be done on the following day should be listed for each worker on the preceding evening. It takes a certain amount of patience to deal in this way with all foreseeable contingencies but the benefits are considerable.

It is also possible to issue blank forms and require workers to fill in the jobs as they proceed from job to job. This practice, however, is not considered very reliable unless the workforce is exceptionally well trained and responsible.

The best procedure in shops employing fifteen or more maintenance men is to appoint a despatcher whose task will be to issue the working copies of the work orders

WORKER'S DAILY TIME CARD				Name............		No............	
Dept/Shop............				Date............			
From	*To*	*Elapsed hrs.* *Regular*	*O/T*	*Job description*	*W/O no.*	*Dept. shop*	*Code*
Totals				Code: L—Lubrication; E—Emergency repair; D—Development; I—Inspection; R—Repair.			
Total hrs today							

Fig. 28. Worker's daily time card showing the hours spent on all jobs during the day.

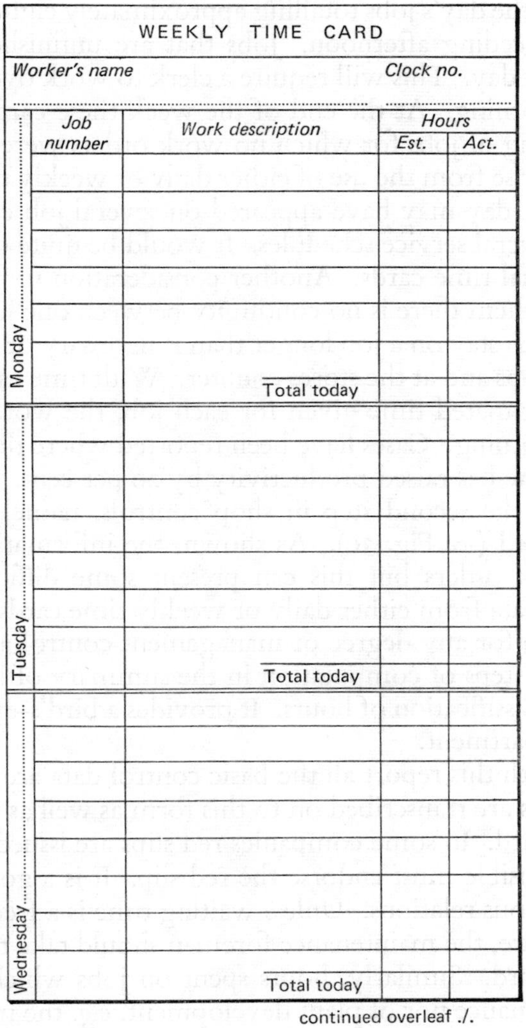
WEEKLY TIME CARD
Worker's name
Clock no.
Job number
Work description
Hours
Est.
Act.
Monday
Total today
Tuesday
Total today
Wednesday
Total today
continued overleaf ./.

Fig. 29. Worker's weekly time card. This provides the basis for craft-hours control.

to each man. When cards of completed jobs are returned and new ones are issued, the despatcher will record the time of completion and starting on both the work order and the worker's time card. Waiting time is also recorded by the despatcher on the time card as it occurs.

Unforeseen jobs of short duration (*i.e.* five to thirty minutes long), can be recorded by the worker himself and initialled later by the foreman. It is advisable to establish this practice so that no new work order has to be issued for these minor jobs and no worker need to return to his foreman's desk to obtain one.

It is also possible to use a weekly time card for the purpose of craft-hours control (Fig. 29). The weekly time card is kept in a rack from which the worker takes it in the morning. Preferably the day's jobs totalling approximately eight hours of work should be posted on the preceding afternoon. Jobs that are unfinished in the evening are transferred to the next day. This will require a clerk to work overtime and prepare the cards for the next morning. At the end of the week these cards are returned for the checking and recording of jobs for which no work orders were issued.

Many advantages arise from the use of either daily or weekly time cards. A worker's time at the end of the day may have appeared on several job cards and a lubricator's time may show on several service schedules. It would be quite difficult to collect these data without individual time cards. Another consideration in favour of time cards is the fact that without them there is no continuity between one job and another, which may lead the worker to stay on a job longer than is necessary or to waste his time both in walking between jobs and at the stores counter. With time cards posted in advance, preferably with an estimated time given for each job, the worker gets his daily plan mapped out in the morning. Cases have been reported where the introduction of time cards in this form alone has raised productivity by 20 per cent.

Now we come to the second step in shop controls, namely the foreman's daily report of hours worked (*see* Fig. 30). As shown, the information could be gathered from individual work orders but this can present some difficulties. It is therefore preferable to get the data from either daily or weekly time cards. The foreman's daily report is indispensable for any degree of management control and if it is well carried out it replaces several steps of computation in the summary of hours for each worker and for jobs and the classification of hours. It provides a bird's-eye view of all activities performed by the department.

It is evident that with this report all the basic control data are easily made available. Hours worked on jobs are transcribed on to this form as well as waiting time from the worker's daily time card. In some companies red slips are issued for waiting time and the department responsible must endorse the red slip. It is a good idea but not really conducive to harmonious relations. Unless waiting time is a frequent occurrence and/or of an excessive nature, the maintenance foreman should take the responsibility for it and initial the time card. Similarly, hours spent on jobs which are obviously not a normal part of maintenance nor of plant development, *e.g.* the making of a prototype, the construction of an exhibition display, etc., will appear in the report. These jobs are time-consuming and it is advisable to keep a critical eye on such requests. The daily summary of hours is a convenient way to do so.

The ease of summarising several horizontal sub-totals and a few vertical ones needs no elaboration but their usefulness is really impressive. By means of this report the control of hours is at one's fingertips. Further detailing and the process of charging hours to departments and to larger jobs can be done separately or by referring to the work order. When data are assembled in this form the later stages of management control follow easily. An additional use of the foreman's daily report appears in Fig. 94, (p. 248) where it serves as the source for data to be plotted on control charts.

The analysis of "clocked" maintenance hours should consist of sub-totals of the sums

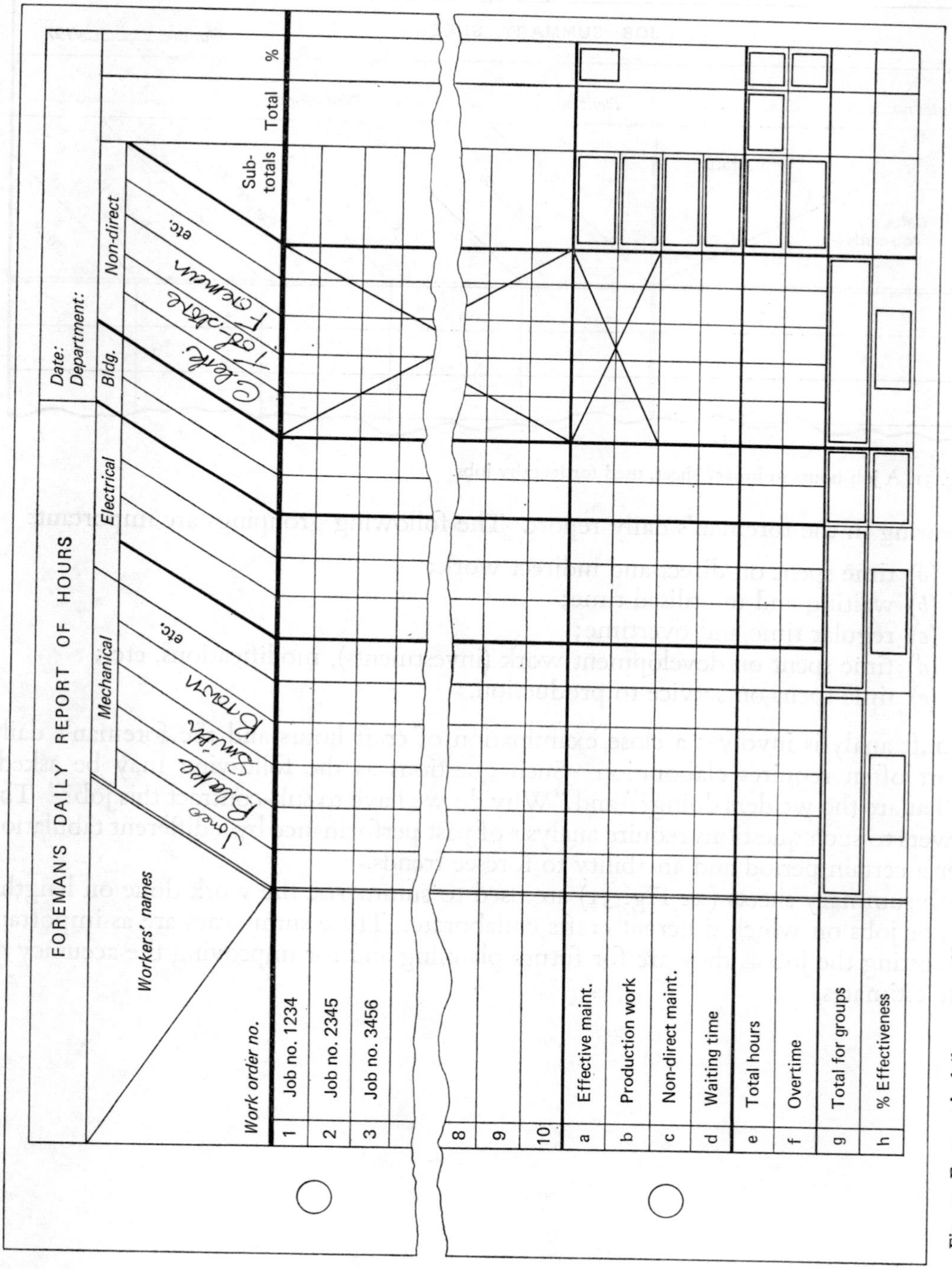

Fig. 30. Foreman's daily report of hours worked.

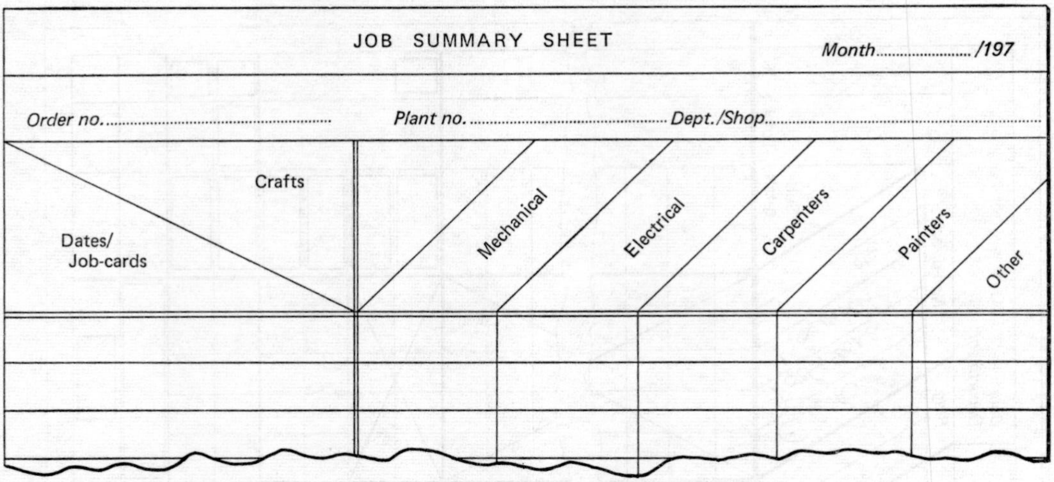

Fig. 31. A job-hours summary sheet, used for lengthy jobs.

appearing in the foreman's daily report. The following groupings are important:

(*a*) time spent on direct and indirect work;
(*b*) waiting and unutilised time;
(*c*) regular time and overtime;
(*d*) time spent on development work (investments), modifications, etc.;
(*e*) time spent on service to production.

Craft analysis involves a close examination of craft hours and the foreman's daily report often requires elaboration. Such questions as the following may be asked: "What are the welders doing?" and "Why do we have to sub-contract this job?" The answers to such questions require analysis of past performance by a different tabulation over a certain period and an ability to foresee trends.

Job summary sheets (*see* Fig. 31) are used to summarise the work done on lengthy jobs or jobs on which different crafts collaborate. These summaries are as important for costing the job as they are for future planning and for improving the accuracy of time estimates.

Chapter 6

Planning and scheduling

The terms "planning" and "scheduling" are often used incorrectly. In the present text *planning* shall be taken to mean that preparatory work which defines *how* a job should be done, by *whom* or by *what* craft it should be done and in what sequence. Similarly, *scheduling* refers to the timing of the job, and *when* and *where* it is to be done, *i.e.* its place along a time-scale according to time estimates or time allocation.

The following section deals with the various types of workload and the means to handle each type in an efficient way. The distribution of the available craft hours is shared among the following:

(*a*) routine schedules;
(*b*) planned maintenance;
(*c*) preventive maintenance;
(*d*) irregular and unforeseen jobs;
(*e*) overhauls, plant shutdowns or project work.

Figure 32 shows the procedures to be adopted in dealing with various workloads of differing degrees of urgency. Once it is realised that types of workload differ, it is easy to set up means to cope with them efficiently. A very frequent reason for a continuous muddle and rush in the department is the lack of advance planning of the last two types. Carrying out such jobs without adequate planning is difficult enough but imposing them on top of regular work will make it still more confusing and, in addition, planned work will suffer as a result.

Many maintenance people claim that scheduling is useless since "Whatever plan you make, it is sure to be upset." This argument can be refuted in several ways, two of which in the form of questions are most effective. The first argument is a negative one: "If you don't plan and don't schedule, could you organise more easily and would you be able to achieve more that way?" The other argument is: "Since we agree that not all jobs can be planned or scheduled, why not do your best with what *can* be scheduled?"

However, before plunging into the problems of planning and scheduling we have to examine some different aspects of the workload, its nature and how to deal with it.

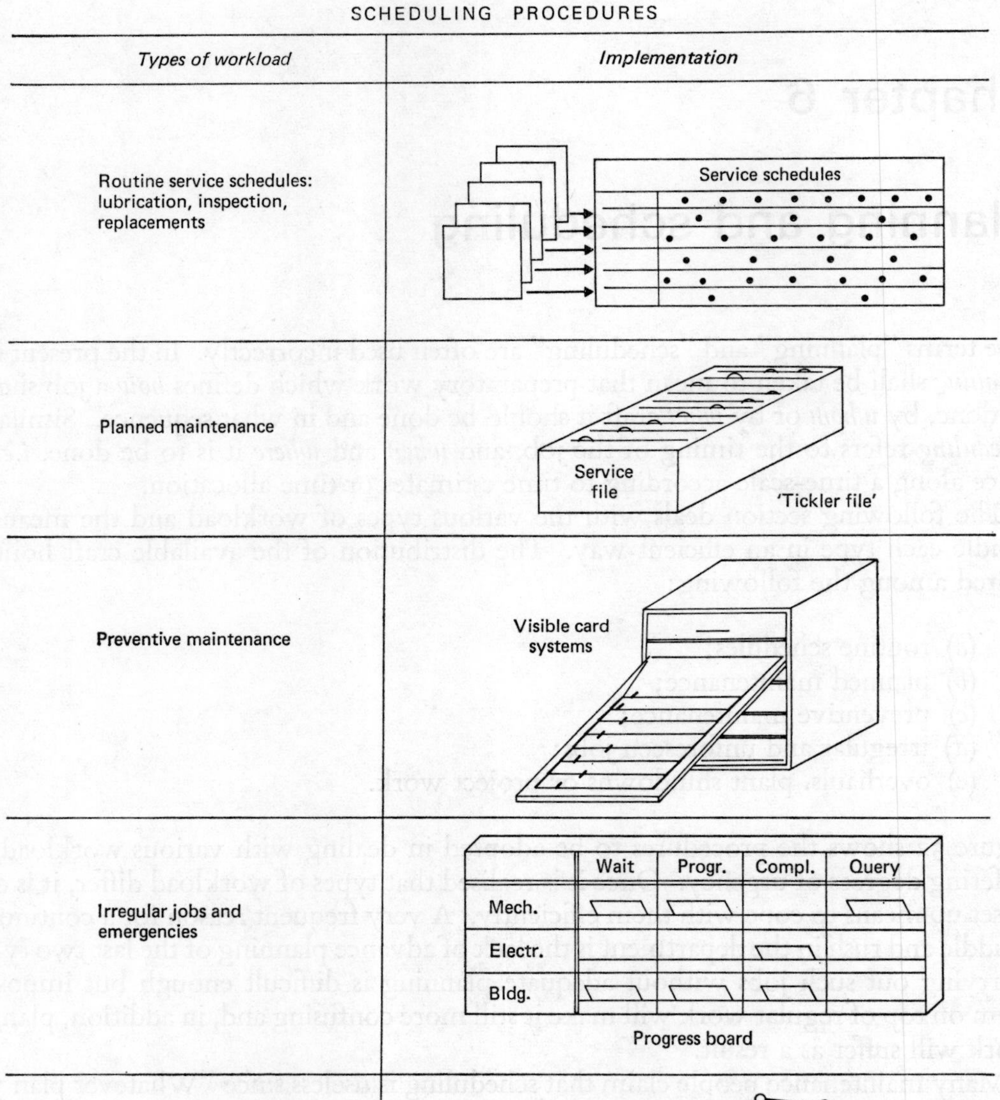

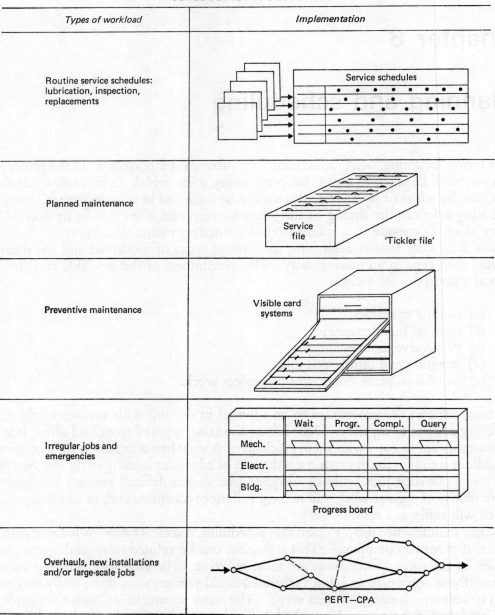

Fig. 32. Different scheduling procedures used for different types of workload.

HOW TO DEAL WITH THE WORKLOAD

The workload situation in the maintenance department is always of direct concern both to management and to the maintenance staff. Is it high or is it low? How much work is there? It is simply rising or is it getting out of hand? What about those jobs for which there never seems enough time?

A systematic answer is found by dividing the problem into several parts:

(*a*) sources of work;
(*b*) keeping track of the workload;
(*c*) striving for an even workload;
(*d*) measures to handle peak loads.

A common fallacy in attempting to handle the workload is to emphasise job planning, whereas more effort should be directed towards *the effective use of craft-time.* Improved sequencing and better timing of services will invariably ease the strain and there are a number of broad principles that can be applied to achieve this.

Firstly, regular and cyclic routine work should be maximised. A homogeneous, even workload will allow workers to accomplish more work with fewer interruptions and more machines will get that frequent attention, revealing those hidden weaknesses which so often lead to breakdowns.

Secondly, permanent assignments must be made for jobs where specialisation is justified. This can be done either by job, *e.g.* band-saw sharpening, steel-cable splicing, etc., or by area. In order to reduce repeated calls to certain areas, excessive walking time and "job-hopping," full-time staff should be allocated to such areas.

Thirdly, the total *regular planned work* should involve between 65 and 75 per cent of the total workforce, leaving the balance for unforeseen jobs and a reserve on jobs that can be interrupted at any time, *e.g.* reconditioning of parts in the central shops. As we have seen in Fig. 11 (p. 47), the total workload decreases as planned work is increased.

Let us now analyse the sources of the workload and the way in which they are dealt with (*see* Fig. 33). Assuming an average, well-run maintenance department we can expect the load to be distributed between five classes of urgency in the proportions shown.

(*a*) *Immediate jobs and emergencies* have to be attended to straightaway, however their incidence decreases sharply as planned work increases. In ideal conditions only accidents and acts of Providence should fall within this group.

(*b*) Quite apart from true emergencies, *urgent jobs* arise either from somebody's neglect of a situation until there is "real trouble," or very often when certain people use their influence with management and maintenance work in their departments is given unjustified priority. Jobs fall into this latter group mostly because of poor communications, *i.e.* maintenance is not notified in advance of certain jobs that have been initiated by management. For instance, if a new machine is to be installed tomorrow, there must have been a chain of events leading to its arrival, from placing

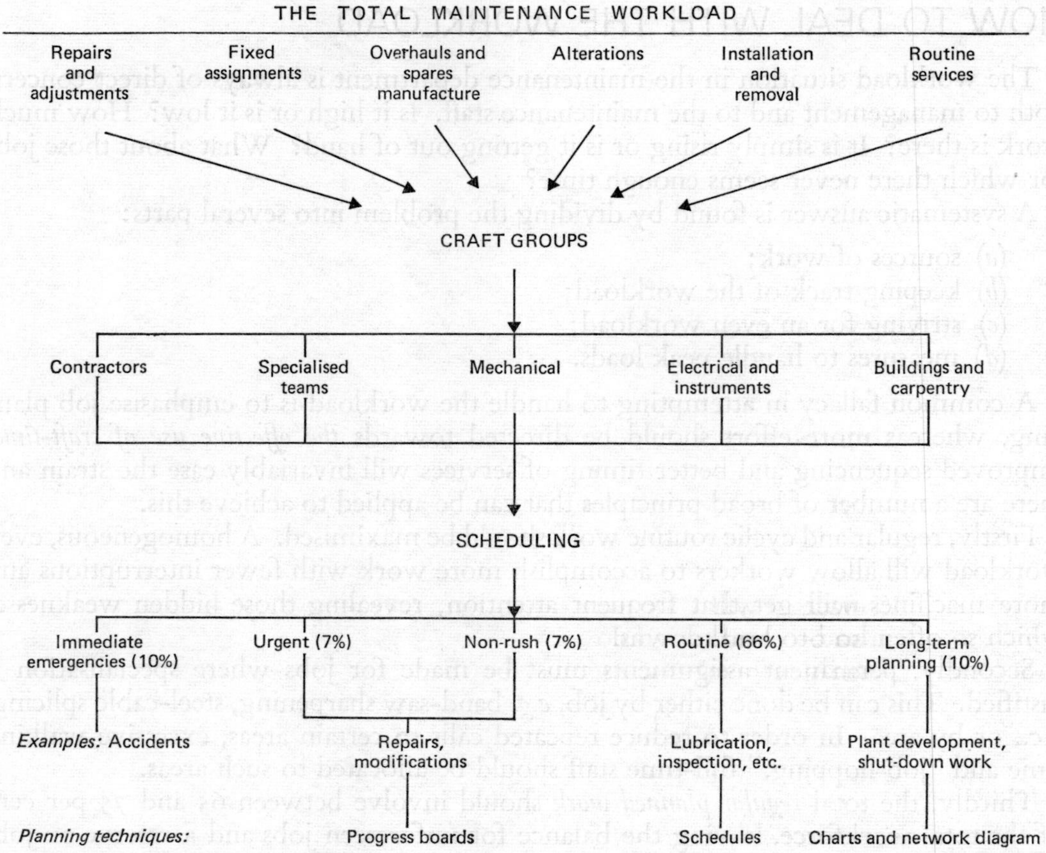

Fig. 33. The analysis and distribution of the workload in one factory and how it is translated into schedules.

the order to obtaining the promise of a delivery date. Or if Department X is to take up its new premises next week the decision must have been preceded by a good deal of discussion. So why not notify maintenance in good time?

(*c*) *Non-rush* jobs are those which management have generously agreed may be fitted in "when you have a free moment." Their number could be further reduced if better relations existed with production supervisors or if an occasional informal chat with management were possible.

(*d*) *Routine schedules* are easy to handle, in fact they should run by themselves without needing any additional attention. Daily, weekly, fortnightly and monthly tasks which recur with a predictable frequency can be timed and accurately estimated. They add up to a certain regular load although on some days the cycle may be more heavily loaded than on others. To even out these fluctuations different jobs can be combined in the schedules, *e.g.* lubrication with adjustments or check-ups. Ideally this type of workload should make up 45 to 50 per cent of the work while another 15 per cent should consist of permanent assignments.

(*e*) *Long-term jobs* such as overhauls or shutdowns or long-term project work must be placed in a schedule beside a yearly scale and distributed evenly. Shutdowns may be seasonal within a one-year or a two-year cycle and since they are unavoidable, unless they are subcontracted, the rest of the workload will have to be adjusted accordingly.

In order to build up the load into manageable form all the jobs falling into the above groups should be listed and tabled. This is where departmental organisation may be affected by the character of the workload. It is no use strictly following some concepts of organisation when the workload could be much better handled if there were some changes made such as permanent assignments to jobs and shops. This will emerge from tabulating the load per area and per craft, both for shops and for departments. For budgeting purposes the plant may be divided into cost centres. A tabulation such as the one proposed in Fig. 34 will assist in defining the major components of the workload.

EXPECTED WORKLOAD (IN HOURS) Period............

Dept/Shop cost centre	Craft group	Routine schedules			Major/minor jobs		Contingencies		Total
		Description	Code	Hours	Description	Hours	Description	Hours	

Fig. 34. Summary of expected workload.

Part of the workload is built up from the routine schedules that we have established. As mentioned earlier these can be measured easily and distributed evenly over an annual scale. On the other hand we can allocate certain amounts of time on a regular basis, such as three half-days a week, four hours per day, etc.

The following example shows a systematic method of planning the workload in the servicing of electric motors.

EXAMPLE: *duty rating of electric motors*[30]

Establishing priorities for servicing electric motors can best be done by rating their conditions of use. To impose a uniform routine on all motors could prove wasteful inasmuch as some heavily used motors require longer and more frequent services. Figure 35 shows a rating system by which the appropriate yearly service frequencies can be set.

The total rating score of each motor will indicate the category into which it will fall.

A service frequency is established for each group of motors (item B in Table VIII) within a certain range of rating points. The servicing procedure involves a combination of past experience and the work prescribed by the manufacturer.

Service instructions are specified for each group and type of service and, naturally, work

	Electric motor duty rating	*Motor no*							
		101	102	103	104	105	106	107	108
Surroundings	1. Humidity and air contamination None — Occasional — Heavy, direct 1 2 3 4 5 6 7 8 9								
	2. Temperature Normal — Fluctuating — Extremely high (or low) 1 2 3 4 5 6								
	3. Ventilation Good — Poor — None 1 2 3 4 5 6								
Load conditions	4. Load Light, occasional — Moderate — Capacity 1 2 3 4 5 6 7 8 9								
	5. Frequency of start and stop Occasional — Moderate — Extreme 1 2 3 4 5 6								
	6. Impact of load Remote and smooth — Indirect and jolting — Direct on shaft 1 2 3 4 5 6 7 8 9								
	7. Duration of running periods Short, occasional — Moderate — Extreme 1 2 3 4 5 6								
Plant factors	8. Importance to plant Small — Considerable — Indispensable 1 2 3 4 5 6								
	9. Availability of stand-by or spares On-the-spot — Feasible — None 1 2 3 4 5 6								
	10. Suitability for job Fully adequate — Marginal — Inadequate 1 2 3 4 5 6 7 8 9								
	Rating total								

Fig. 35. Duty rating sheet for electric motors. For each motor the rating is recorded in the appropriate column and totalled. The total rating score of each motor will indicate the category in Table VIII into which it will fall. (This illustration was first published in *Factory*, McGraw–Hill, New York, June 1962.)

content will also vary according to the size of the unit. In broad terms the check-up includes:

(*a*) visual check and electrical connections;
(*b*) alignment, vibration and bearings;
(*c*) insulation tests and overheating;
(*d*) overhaul—when indicated by inspection.

By estimating service time, materials and their corresponding costs the annual total can be calculated (*see* Table VIII).

Table VIII. Example of annual workload and budget for a group of electric motors; an example showing the conversion of workload (A, B, C) into time and cost estimate (E and G).

Range of rating (points)	5-21	22-39	40-59	60-90	*Totals*
A Motors in group	23	49	52	31	155
B Services per year each motor	2	3	4	6	
C Total services per group (A x B)	46	147	208	186	
D Average time per service hrs	0.1	0.15	0.2	0.2	
E Annual total time per group (C x B) hrs	4.6	22.1	41.6	37.2	105.5 hrs
F Cost per service (£)	£1.00	£1.30	£1.40	£2.00	
G Total annual cost for group (C x F) (£)	£45.00	£200.00	£380.00	£400.00	£1,025.00

The hours required to perform the prescribed services must now be distributed along a time-scale. When a large number of such services are to be scheduled, the problem arises of distributing the workload evenly over a given period. Figure 36 shows an example where the schedule form can assist in this procedure.

In this case a planned service schedule is illustrated. Let us assume that we have assessed the times required for each service and that we have placed them along a time-scale. Depending on the frequency, type of service and number of similar machines, we obtain a varying load for each week. In trying to even out the load we have either to stagger the services or break them up into groups of similar units. The form allows several trials to be made so that the sum of hours required for these services each week should present approximately an even load. We may also deliberately want to obtain some days or weeks with no load at all. Other assignments will then have to be found for those intervals such as overhauls or reconditioning of components.

Whether the posted times are assigned minutes or target times is a matter of individual preference. Whatever the choice the approach has to be uniform throughout all maintenance activities.

Once the regular jobs (schedules and permanent assignments) are recorded and scheduled the major part of the workload becomes clear. This may amount to about 50 to 60 per cent or more of the available man-hours. Next it is advisable to install a procedure for keeping track of the fluctuating load imposed by emergencies, repairs,

Dept. A ANNUAL SCHEDULE OF MECHANICAL SERVICING OF MACHINE TOOLS Year: 197

Machines	Services			Week by week service time (minutes)																													Total time per year		
	Periods	Min.	Freq.	1	2	3	4	5	6	7	8	9	10	11	12	13	14	15	16	17	18	19	20	21	22	23	24	25	26	27	51	52	One m/c	X m/cs	Hours
One brake press	Weekly	5.00	26X		5		5		5		5		5		5		5		5		5		5		5		5		5			5	130.00		2.2
	2-weekly	8.00	13			8				8				8				8				8				8							104.00		1.8
	Monthly	10.00	9					10				10								10				10				10					90.00		1.5
	3-monthly	12.00	2X													12																	27.00		0.4
	6-monthly	·15.00	1X																											15			15.00		0.2
	Annually	30.00	1X	30																													30.00		0.5
	Total annual services		52																														Total	hrs	6.6
Ten drill presses	2-weekly	4.00	13X				40				40				40				40				40				40							440.00	7.6
	Monthly	6.00	6X						60								60								60									360.00	6.0
	2-monthly	8.00	5X										80								80													400.00	6.7
	6-monthly	10.00	1X																											60				100.00	2.2
	Annually	15.00	1X		150																													150.00	2.5
			26																														Total	hrs	25.0
Four turret lathes	Monthly	8.00	6X							32								32								32								192.00	3.2
	2-monthly	10.00	5X											40								40												200.00	3.4
	6-monthly	14.00	1X																											56				56.00	0.9
	Annually	20.00	1X			80																												80.00	1.4
			13																														Total	hrs	9.3
1st trial	Weekly totals (min.)			30	155	88	45	10	65	40	45	10	85	48	45	12	65	40	45	10	85	48	45	10	65	40	45	10	5						
2nd trial	"	"	"																																
3rd trial	"	"	"																																

Remarks:
(a) In this example no attempt is made to level the weekly load
(b) Easter and Xmas weeks may have to be deleted
(c) Annual shutdowns may affect this schedule
(d) This schedule relates to one craft only

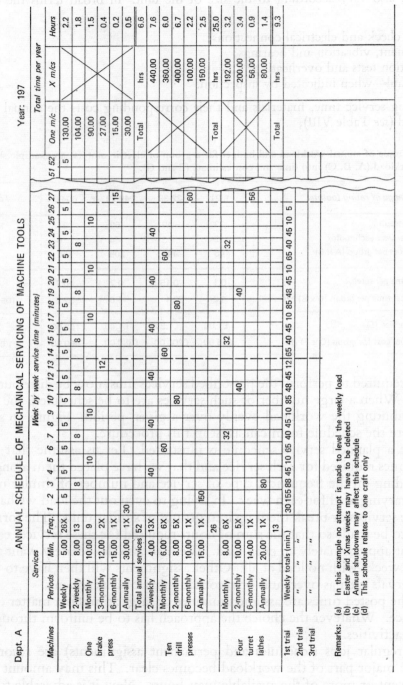

Fig. 36. A method for obtaining an even distribution of the workload on routine services.

overhauls and development (capital) work of one kind or another. Such a procedure is illustrated in Fig. 37. Incoming jobs are posted on a log sheet. Even a rudimentary estimate of hours on each of these will allow them to be summarised weekly. It is also easy to transcribe the total hours completed on these jobs from the foreman's daily report. The total of completed jobs reduces the workload and leaves a balance that represents the workload at the beginning of the next week.

How to handle "overload"

The problem of overload should be approached selectively by identifying items that should take priority and be given preferential treatment. This may throw new light on the problem and on its solution.

Lest we jump to the premature conclusion that more men are needed in the department, let us examine other ways of handling excessive accumulated workloads. Some time-saving devices are discussed in Chapter 8, under "Work study in maintenance," but the following alternatives should also be considered.

(*a*) An improvement of the use of craft hours, through more detailed planning, by advance preparation of parts, tools and materials by storekeepers or apprentices. This will lead to minimising waiting and travel time of craftsmen. Employment of a job despatcher will also contribute towards the same end.

(*b*) The designing out of weaknesses traced from the results of breakdown analysis—*i.e.* the corrective maintenance approach—eliminating, where feasible, the need for servicing.

(*c*) Giving each job request a second look and asking: "Is it really necessary?" Some jobs, requested in a fit of enthusiasm, may not be necessary after all if maintenance can state with accuracy: "The whole process is due to be replaced, you know!"

(*d*) "Let's work overtime," is easy enough to say. But overtime is as inefficient as it is expensive and is to be avoided where possible, and only allowed when the pressure of work demands it (the foolish practice of "awarding overtime" in lieu of installing a bonus scheme is ignored here).

(*e*) Subcontract—in some cases this may be both feasible and economical but experts warn that *occasional* contract work is not economical, compared to regular long-term contract work. It is sometimes argued that if we can plan for outsiders why can't we plan for our own men? Specialised contract work, however, is advisable.

(*f*) More frequent shutdowns may be the answer. However, when we arrive at this conclusion it is indicative of the poor condition of the equipment and other measures may be necessary.

(*g*) Increase the frequency of inspection and lubrication. This is an added workload which nevertheless pays for itself.

(*h*) Pass on some of the simpler routine services to production operators. This measure is sometimes feasible but one should beware of misunderstandings and forgetfulness.

(*i*) Improve stock of spares and stand-by equipment and replace certain parts more

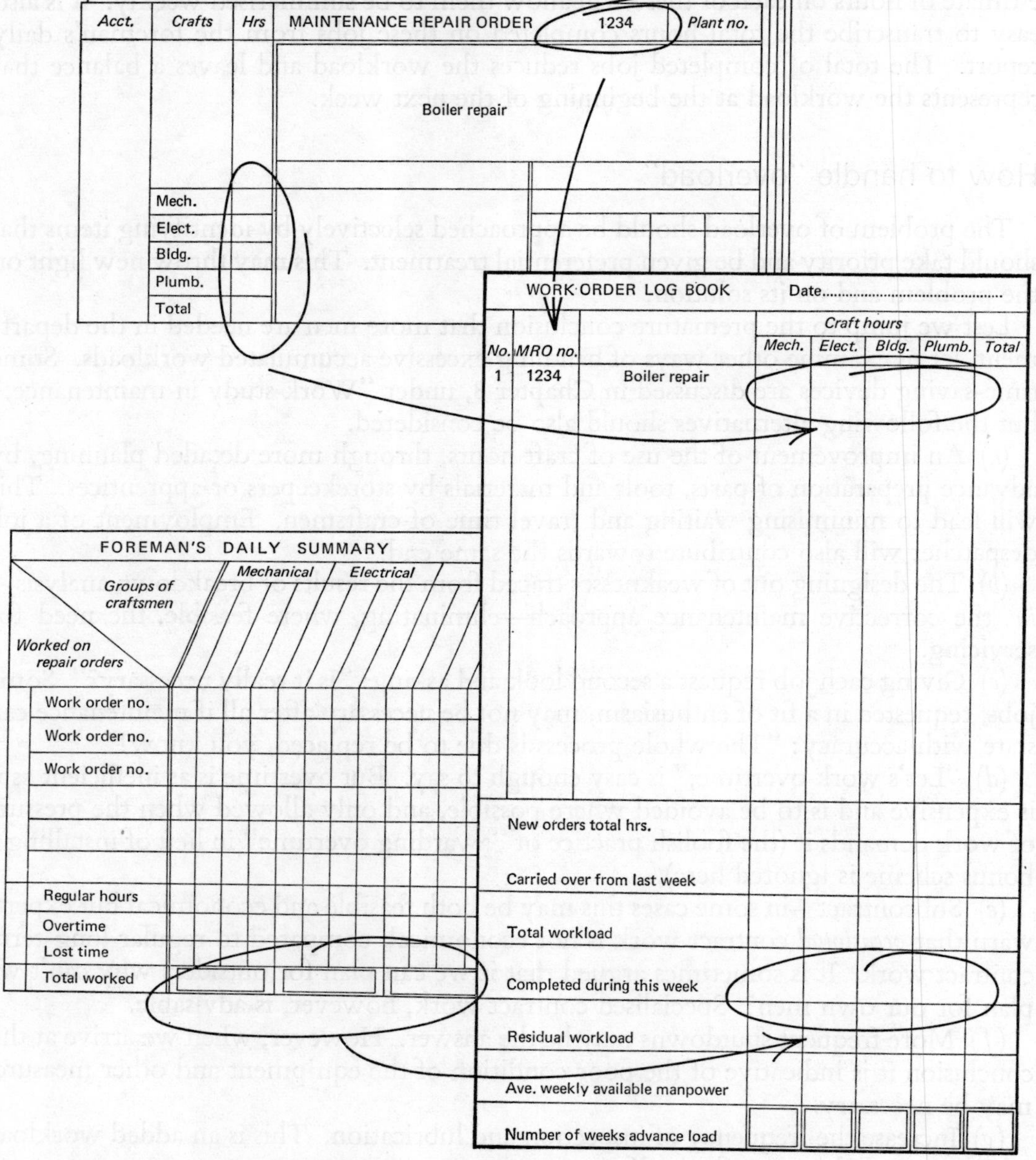

Fig. 37. Procedure for keeping a workload log-book up to date.

frequently, at any rate before major breakdowns occur. This is a preventive maintenance principle.

(*j*) Up-grade craft skills and improve motivation through training. This measure *always* bears fruit and is often the most neglected one, especially in a period when the workload is heavy.

(*k*) Improve communications with production functions and increase the ability of production supervisors to identify failures and to anticipate them.

(*l*) Apply method-study techniques for all major activities. This tool has a powerful potential. It shows up areas where craft time is wasted due to poor working methods (*see* Chapter 8).

(*m*) Apply work measurement with the obvious objective of introducing incentives (and particularly for some of the less glamorous jobs, *e.g.* cost control, planning and scheduling and teamwork balancing). The potential of incentive schemes should not be underrated. However, total benefits to be derived have been the subject of controversy among experts for two decades. In any case, this is a long-term solution and hasty action is to be avoided.

(*n*) Use analytical techniques, such as work sampling and breakdown analysis. These tools usually produce revealing results (*see* Chapter 8).

(*o*) Review the plant periodically, enquiringly and critically.

It should be borne in mind, however, that the more a maintenance team is overworked the less they are able to absorb some of these corrective measures. The case of a small detergent factory provides an appropriate example.

A consultant was requested to suggest ways and means of overcoming the backlog of workload and the delay in the implementation of planned maintenance. During his visit planning procedures were surveyed as well as the repair teams at work. Little seemed amiss until the workload situation was examined. The log book where incoming requests were recorded showed the jobs which were as yet incomplete. It was found that there were not many but that they involved some of the best craftsmen. It showed among other things that the installation of a new process had been going on for many weeks and that the extension to the stores was still in progress. Since these jobs lay in the fields of plant development and investment, it was queried whether they could be subcontracted. It was also shown in a cursory calculation that 30 per cent of the total crew was on this type of job. This had not been realised and thus the reason for delays in regular maintenance was identified and the problem eventually solved. A situation which appeared as excessive workload in fact resulted from a wrong assessment of priorities.

ROUTINE SERVICE SCHEDULES

As previously described in Chapter 4, in "The four systems," routine schedules are characterised by a regular repetitive cycle that depends basically on what can be conveniently done within a given period. The rate of progress of a routine schedule can be determined by the amount of time spent per day, per week or per month. With a sequence of steps for the service-man established, the routine proceeds on its course, practically by itself. As a follow-up it is only necessary to enquire whether progress was as forecast, and whether anything has occurred to hold it up.

Routine schedules can best be explained by the following three cases. The first has been chosen to illustrate what can be obtained by introducing very simple routine services.

CASE I: *Routine servicing of tool-grinders, using a Type C form*

A certain machine shop, containing about fifty machine tools, had about ten conventional tool-grinders distributed between machines. The shop supervisor noticed that queues of operators or machine-setters tended to build up next to certain of the tool-grinders, giving the waiting men a great opportunity for playing around. The tool-setters complained that as a result they could not work satisfactorily.

A survey showed that the grinders in question, being the responsibility of no one in particular had been badly neglected and were in poor condition. Queues used to collect at the more useable onces. The situation had become a source of real irritation to everyone. The remedy for this state of affairs was the application of a regular maintenance procedure.

A single form was drawn up where all possible defects were pre-printed and grouped according to part, as illustrated in Fig. 38. A small supply of forms was attached by means of "photo-corners," to a clipboard (Fig. 39) with a pencil attached and hung on a nail next to the maintenance foreman's desk. Instructions were issued to the effect that starting on the following Monday all tool-grinders would be inspected daily and the forms marked accordingly. Subsequently all repairs were to be carried out promptly. The foreman was instructed that at the end of the day he should present the results duly recorded on the form to the plant engineer.

On the Monday morning in question the foreman assigned one of his mechanics to this job and the latter proceeded to "fix" the tool-grinders. It was a tedious task, giving rise to irritations and frustration. By mid-afternoon he had a number of tool-grinders in good condition. On some of them the grinding-stones were found to be down to the shaft, and in some guard-housings the abrasive, mixed with chips, had combined into a hardened mass which needed chisel-and-hammer work to dislodge it. The electrician had to be called on several occasions to repair the switches and some tool-grinders had to be dismantled for the bearings to be replaced. Some jobs could not be completed on the same day because parts were not available. The task had so far taken a day's work.

The first completed form was grimy and greasy, but it was initialled and filed. The incomplete jobs were transferred to the next form and the mechanic was instructed to complete them on the following day. By noon next day most of the tool-grinders were working again.

The maintenance foreman now issued the instruction that until further orders all grinders were to be inspected and repaired twice a week, so *two* completed forms per week were to be expected. This routine was kept up for three weeks when it became obvious that *one weekly* check would suffice. Instead of checking each item on each machine the mechanic could now check a whole machine and mark his work with one tick. From that time onwards the once a week check on the ten tool-grinders took fifteen minutes.

The effect on the shop was tremendous. Whenever anybody needed a tool-grinder, it

WEEKLY INSPECTION AND REPAIR OF TOOL GRINDERS

Account no.

Hours..........Stnd. min..............

Total time allowed for 12 m/cs

	Grinders no.	1		2		3		4		5		6		7		8		9		10		11		12	
	Instructions	CH	R	CH	R	CH	R	CH	R	CH	R	CH	R	CH	R	CH	R	CH	R	CH	R	CH	R	CH	R
Mechanical	Grinding stone																								
	Guard																								
	Tool-rest																								
	Bearings																								
	Lubrication																								
	etc.																								
Electrical	Main switch																								
	Light switch																								
	Motor																								
	etc.																								
Accessories	Goggles																								
	Coolant																								
	Chip guard																								
	etc.																								
Appearance	Cleanliness																								
	Vibration																								
	etc.																								
	Date																								
	Signature																								

Fig. 38. Weekly routine inspection sheet for tool-grinders.

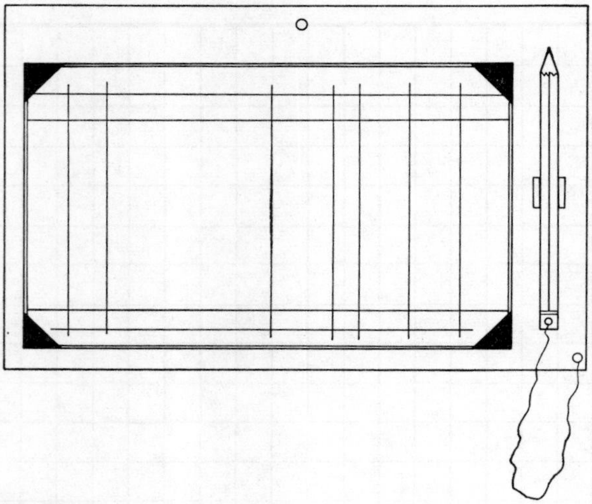

Fig. 39. Clipboard-mounted forms.

was in good working condition, with water in the container, goggles on the hook, lighting switch working, etc. The queueing was eliminated and tempers had no occasion to rise. All this was maintained at the weekly cost of fifteen minutes' work.

Four points must here be emphasised. Firstly, if the instruction had been given to Jones to "keep all these grinders in good working order at all times," results would not have been as good. Instructions must be specific, clearly understandable and concisely pre-printed.

Secondly, recording and a signature were required, and the completed form had to be submitted for inspection and filing. If these points had been left out, the system would very soon have degenerated and finally collapsed. When nobody shows an interest in results this is bound to happen.

Thirdly, everything needed was attached to the clipboard allowing no excuse for lack of forms, of a pencil or somewhere to write. A bunch of loose forms would have been stuffed into a drawer and forgotten about.

Fourthly, the principle of grouping machines was followed, since the tool-grinders required similar servicing, regardless of the differences in size and make. In this manner all critical groups such as pumps, starters, exhausters, etc., can be handled. Once the system is in operation problems regarding that particular group are solved.

Instructions such as these are easy to follow once a regular pattern has been established, and if the form is called "weekly inspection and repair" no additional initiating action is required. The form serves as (*i*) a *reminder*, (*ii*) a *checklist*, (*iii*) a *work order*, (*iv*) a *permanent record* and (*v*) *evidence*, in case bad workmanship comes to light and disciplinary action has to be taken.

Figure 40 shows a simple form used for inspections of hotel rooms. This is also an example of a routine and *not* a preventive maintenance procedure, since no measures at all are taken to prevent the occurrence of defects. In addition, it is not a case of planned maintenance where the servicing dates are fixed in advance or carried out according to machine usage.

CASE II: *Inspection of exhaust fans, using a Type D form*

This case concerns four exhaust fans, two of which were operating in spray-booths and two in a woodworking-shop. It had been erroneously assumed that these fans would work indefinitely without attention, until difficulties started arising with increasing frequency. It was decided that quarterly inspections coupled with adjustments and repairs would be adequate. A form which would service the four fans was drawn up accordingly as in Fig. 41.

The form was issued, and in January the four units were satisfactorily serviced. The maintenance clerk was instructed to note in his diary the service dates due for the rest of the year, and he was instructed to issue the same form on these occasions. In this manner *one* form took care of four services for a whole year, and it solved this particular problem once and for all. The form was subsequently issued three times during the remainder of that year.

The grouping of similar units is particularly useful even when units are not completely identical, since the names of components are common. It is here suggested that whenever groups of units exist, their service should be tackled on a similar basis.

Scheduling problems arise when, with variety in types of equipment and the sheer number of services required, the amount of work increases. The workload arising out of a variety of services, spaced at different intervals and requiring different crafts, necessitates a more complicated kind of paperwork. This is illustrated in the next case.

INTERNATIONAL HOTELS ROOM INSPECTION CHECKLIST

HOUSEKEEPER

ITEMS TO BE CHECKED

✓ - check-mark for O.K.
x - requires attention
xx - immediate repair completed

		ELECTRICAL				BATHROOM				FURNITURE				HOUSEHOLD			
Room no.	Date	Lights and bulbs, switches	Air-conditioner	Refrigerator		Bath and fixtures	Toilet and flushing	Washbasin	Towels, paper, soap	Carpet and curtains	Walls and pictures	Beds and chairs		Menu and directory	Stationery	Dust and cleanliness	
401																	
402																	
403																	
404																	
411																	
412																	
413																	
414																	
415																	
416																	
417																	

Fig. 40. Simple inspection checklist for hotel rooms.

QUARTERLY INSPECTION OF EXHAUST FANS											Acct. no.		Total time allowed				
Instructions \ Machine		Exhauster 1				Exhauster 2				Exhauster 3				Exhauster 4			
		1st	2nd	3rd	4th	1st	2nd	3rd	4th	1st	2nd	3rd	4th	1st	2nd	3rd	4th
Mechanical	Drive pulleys Vee belts Bearing Lubrication etc.																
Electrical	Main starter Motor temperature etc.																
Miscellaneous	Ducts Cleanliness Noise and vibration etc.																
	Date																
	Signature																

Fig. 41. Quarterly routine inspection of exhaust fans.

JIG BORERS 3

LATHES 5

DRILL PRESSES											
Frequency of service								Number of machines			7
Plant no. / Instructions	Code	Freq.	Time per service	Annual time	Plant no. 101	102	103	104	105	106	107
Lubrication:	△	3x									
Mechanical Service:	○	4x									
	✱	6x									
Electrical Service:	□	2x									
			Date								
			Signature								

(Type C form)

COMPREHENSIVE ANNUAL ROUTINE SCHEDULE MACHINE TOOLS

CODES: △ - Mechanical; ○ - Electrical; ✱ - Lubrication; □ - Inspection

Machines / Services	Plant no.	WEEKS 1	2	3	4	5	6	7	8	9	10	11	12	13	14	15	16	17	18	19	etc.
Drill	101	△	✱				○														
presses	102	△	✱				○														
	103	△	✱				○														
	104	△	✱				○														
3x yearly	105	△	✱				○														
	106	△	✱				○														
	107	△	✱				○														
Lathes	201			□	△			○													
	202			□	△			○													
	203		□		△			○													
6x yearly	204		□		△			○													
	205	□			△			○													
Jig borers						○															
etc.																					

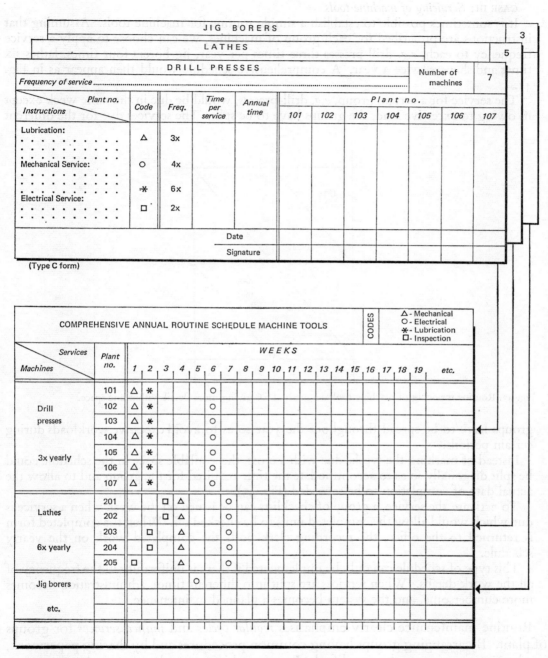

Fig. 42. Routine servicing of machine tools. Planning and recording by groups of machines is more efficient.

CASE III: *Servicing of machine tools*

It is sometimes possible to establish a simple routine for machine tools. Assuming that instructions are prescribed for each group of machines, we can choose to apply a service frequency to each, *e.g.* drill-presses three times per year, jig-borers four times, lathes six times and shapers twice a year. A comprehensive schedule would then appear as in Fig. 42.

The service for a certain group, *e.g.* drill-presses, would be due in the same week except if, owing to excessive load, the groups have to be split. The services due for the different

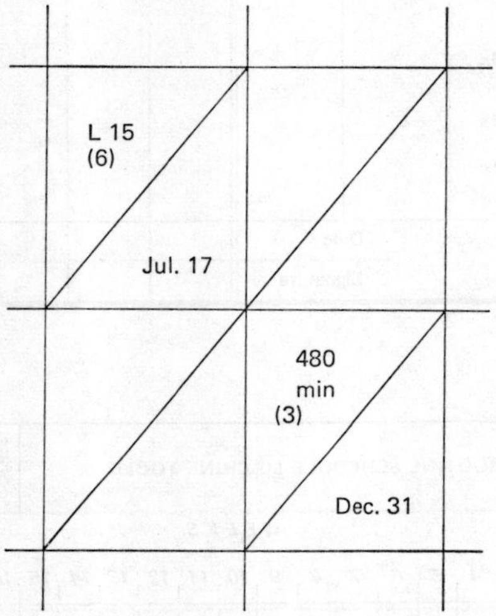

Fig. 43. Routine servicing of machine tools—a method for giving more detail within one space.

groups have to be spread throughout the year so as to avoid excessive workloads during certain periods.

Instead of marking the weeks due with a cross, the available space in the schedule could be split diagonally so as to accommodate the time required for the group and to allow the actual date of completion to be recorded (Fig. 43).

To activate the service, a group of checklists can be issued on the week when a service is due which would allow the completed units to be checked off. When the completed form is returned to the office, the recording clerk posts the completed work on the yearly schedule.

This type of schedule can only be implemented if there is uniformity in the frequency of all the work details. When services are split into different times, administration becomes more cumbersome, and the system becomes a planned maintenance.

Routine maintenance clearly emphasises *regular cycles* and *unified services* for groups of plant. By arranging groups having common procedures and by sharing paperwork, implementation is greatly simplified. In using this approach, not much accuracy is sacrificed and in most cases service is just as effective as other more costly systems.

SCHEDULING OF PLANNED MAINTENANCE

Planned maintenance is based on servicing periods and instructions which are prescribed in much greater detail than in the preceding system. These are affected by the age and condition of the equipment, and its utilisation, and by the needs of its working parts. This approach is markedly different from routine schedules, where the main consideration stems from yearly frequencies that are both feasible and practical and thought to be the most economical and effective. In broad terms, planned maintenance takes care of details that routine maintenance will ignore for the sake of simplicity. In the routine system, equipment tends to be handled in groups for the sake of expediency while in planned maintenance the service is "tailor-made" for individual machines.

Obviously, then, planned maintenance is more elaborate, more painstaking and, for that matter, more costly and cumbersome to administer. We therefore turn our attention to the means by which we can best operate the system economically and at the same time follow the life of individual machines in detail.

Planned maintenance should start when the plant is purchased and installed. At this initial stage much can be done to prepare for and facilitate subsequent maintenance by mapping out the life of the unit. Installation practices are often hasty and therefore poorly planned. In one instance, owing to a certain sequence of installation, it was overlooked that a draw-out switchgear unit required an electrician to stand next to it during maintenance. Since this had not been foreseen the only way servicing could be done was by means of a raised platform which could straddle the unit from above. Many readers will certainly have had similar experiences of their own.

All plant units should have a projected life-span, an expected period between overhauls and an expected schedule for replacing some of their components. Some of this information, often somewhat optimistically biased, will be provided by the manufacturer. But unless we have gained our own experience by making a comparative assessment with similar units, we must rely for a start on what the manual tells us. Evidently what has to be done will differ according to the nature of the equipment, its size, its complexity and its usage.

A different problem is that of assuming responsibility over a plant which has been in operation for some time, and where perhaps maintenance had not been good in the past. This is not an easy task. Undoing the effects of neglect, retrieving soiled manuals from tool-chests and collecting widely dispersed spares from odd corners, can be frustrating. In such cases we often have to arrive at decisions according to the best of our abilities and start from scratch.

When planned maintenance is being introduced, an internal "sales campaign" may be helpful. Often production supervisors and management have to be won over to a certain idea, occasionally even some of the maintenance staff may need some convincing. The fact that planned maintenance takes a long time to get under way may be conceded initially. But all too often confidence fades, people start getting impatient and management may withdraw its support. Such a course may be avoided by a sort of publicity campaign, perhaps in the form of briefing sessions. It may also be helpful to introduce routine servicing on groups of items which can be operational sooner so that some tangible results can be witnessed by everyone at an early date. It is to be hoped that not everyone has had unhappy experiences that has made them sceptical of this kind of publicity. A certain amount of flag-waving to promote better understanding is therefore to be recommended.

Getting the facts together for a thorough maintenance plan can be tedious. Extracting the information from the service manual means transcribing it on to our own forms.

ANNUAL PLANNED MAINTENANCE SCHEDULE

Plant no.	*Service code no.*	*Service*	*Annual freq.*	*Time per service*	*Total min.*	*Weeks* 1	2	3	4	5	6	7	8	9	51	52
Machine plant no. 101	L.1	Lubrication A	12	15	180											
	L.3	Lubrication B	6	20	120											
	I.5	Inspection	2	10	20											
	H.1	Overhaul	1	960	960											
	R.4	Replacements	1	30	30											
Conveyor no. 305	Pa.2	Paintwork	2	180	360											
	Pi.1	Pipe fitting	2	25	50											
	E.3	Electrical	2	10	20											

Fig. 44. Annual planned maintenance schedule.

The design of forms that can serve a variety of equipment and its needs is not easy either. It has to be borne in mind that the basic objective is to put instructions into workable form. Depending on the nature of the equipment, we may adopt a certain implementation sequence, proceeding by groups, by departments or by processes.

It is perhaps best to try to visualise the final form of a schedule as shown in Fig. 44 and to work towards it, *i.e.* working from the end result backwards.

The services required for each machine must be properly documented. Every plant unit will probably have a set of instructions consisting of separate sheets for each craft group dealing with the regularly recurring planned services. It is best to divide these so that they conform to the organisational grouping. For instance, in any medium-sized plant there will be a lubrication crew and perhaps an inspection team, whether part-time or full-time. In larger enterprises there will be an overhaul team or the central mechanical pool will be capable of overhauling machines. This being the case, instructions will be listed on separate sheets as in Fig. 45, and each team will receive the service sheets relating to its particular tasks. Each form is subdivided into a rising

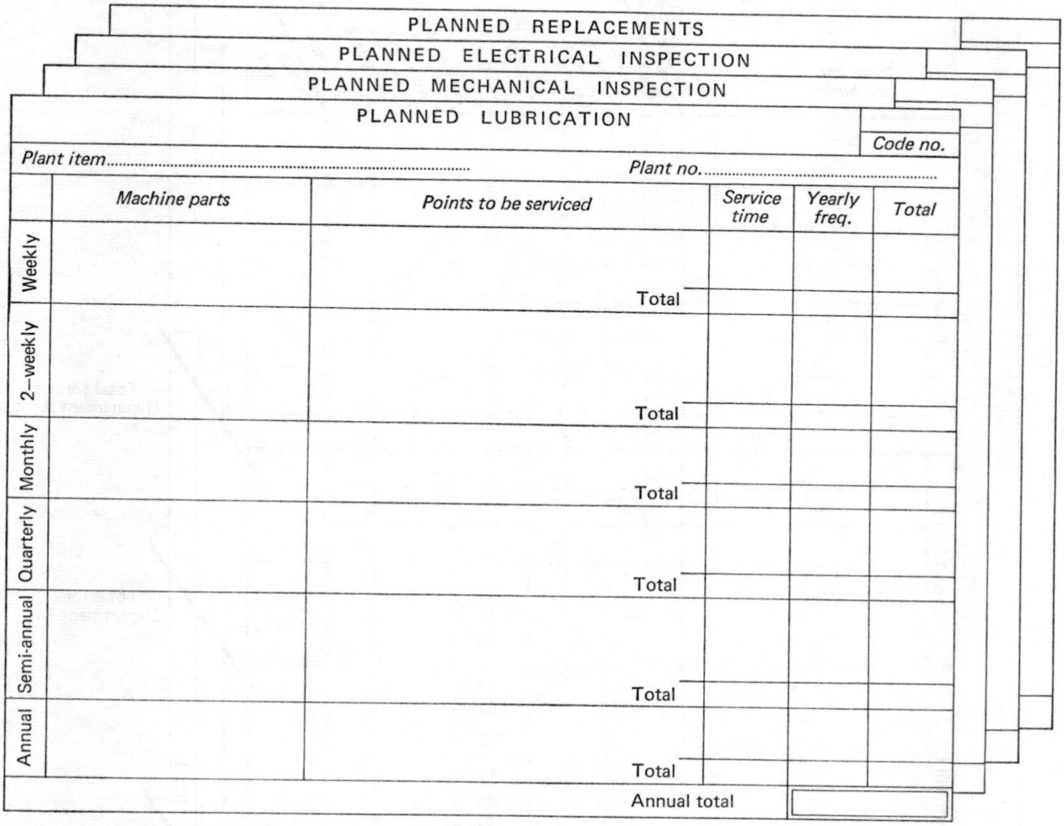
PLANNED REPLACEMENTS
PLANNED ELECTRICAL INSPECTION
PLANNED MECHANICAL INSPECTION
PLANNED LUBRICATION
Code no.
Plant item
Plant no.
Machine parts
Points to be serviced
Service time
Yearly freq.
Total
Weekly
2–weekly
Monthly
Quarterly
Semi-annual
Annual
Total
Total
Total
Total
Total
Total
Annual total

Fig. 45. Listing of service schedules by frequency.

sequence of frequencies. The total time for each service can be filled in, and by multiplying service time by the projected frequencies the total annual load created by every plant unit for each team can be calculated.

Since a subdivision of the lubrication workload is by department, an additional sheet (as in Fig. 46) can be used to total all the services assigned to a department. These forms carry the totals by frequencies and crafts. So having documented and grouped these services, the next stage is the preparation of semi-annual or annual schedules.

At this stage it is important to make certain decisions concerning the schedules. Each schedule can handle a certain number of plant units or machines. But do we prepare lubrication schedules and inspection schedules separately or are we to combine them?

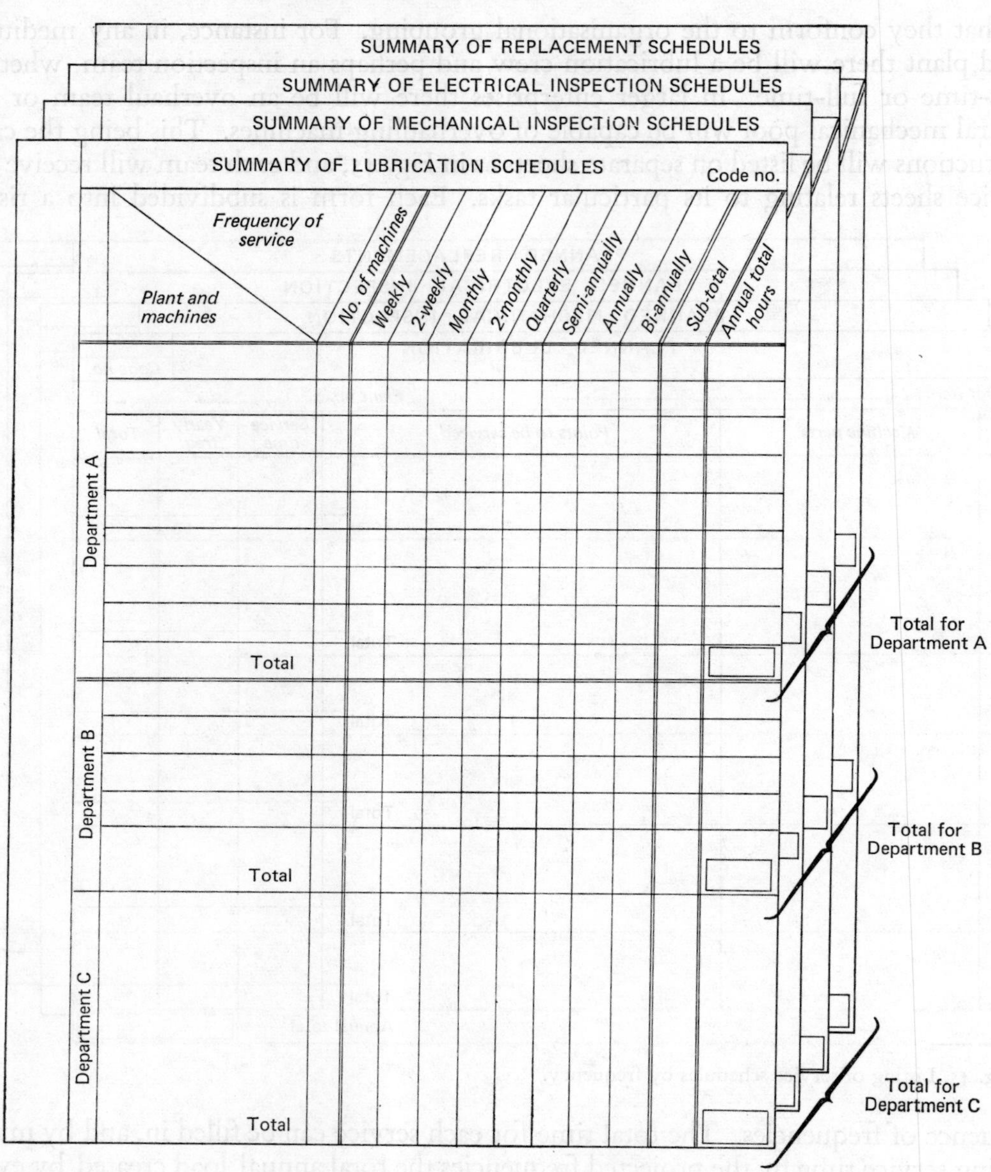

Fig. 46. Forms showing totals of all services assigned to a department, including frequency of service, craft hours and type of work.

Shall we group machine tools and transport equipment in separate schedules or should schedules relate to areas, shops or production departments?

Whether we decide on one course or another there must be a good reason for our decision. In broad terms, a departmental grouping (or grouping by areas) will be suitable for large enterprises. Schedules by groups of equipment are also attractive since they are conducive to efficiency by serving one homogeneous group, the tools and spares used being similar. The same argument speaks in favour of craft groupings, although this division should only be maintained if the service groups in fact function independently of each other. As already mentioned, some services may be combined so as to reduce the frequency of attendance on individual machines.

Coding is used to indicate the various services. For each machine there is a set of service instruction sheets, namely: lubrication, inspection, overhauls, etc., which are identified by plant unit number and followed by a service code. If there are several identical machines, their service instructions will bear the same code number which will serve the whole group.

It is convenient to allocate a code number that indicates the frequency of the service, the craft involved and the main purpose of the service (*see* Table IX).

Table IX. Coding for service instruction sheets.

Annual frequency		*Purpose*		*Crafts*	
Weekly	– W	Lubrication	– L	Mechanical	– Me
Two-weekly	– 2W	Adjustment	– D	Electrical	– El
Monthly	– M	Inspection	– I	Pipe fitting	– Pi
Quarterly	– Q	Replacement	– R	Paintwork	– Pa
Semi-annually	– 6M	Cleaning	– C	Carpentry	– Ca
Annual	– A	etc.		Transport	– Tr

The summary sheet for each machine as shown in Fig. 45 will allow calculation of the annual workload and enable the establishment of an effective servicing routine. For instance, by placing the various services pertaining to one machine side by side it is easy to discover and reduce excessive service frequencies by grouping them into a minimum number of trips.

Equally important is the fact that there is accurate evidence for the planning of manpower, the revision of frequencies and the costing of these services, *making sure that they are covered by job orders*. This approach therefore facilitates budgeting and cost controls.

It now becomes clear that planned maintenance aspires to much higher goals of accuracy and thoroughness than routine servicing. It is also correspondingly more expensive. More organisational effort is needed, more clerical work is required at the outset and more work is necessary to keep it running. However, since the activity round each plant unit multiplies as the frequency and thoroughness of attention rises there is little time for serious defects to develop and thus the objective of breakdown prevention is very nearly achieved.

To keep costs at a reasonable level we must employ multi-craft mechanics and a well-qualified planner who can constantly be on the alert against duplication of effort and "over-maintaining" of plant. For instance, if excessive service frequencies imposed at the start later cause a heavy clerical load to build up both in the issuing of instructions and at the time when they are recorded, then the planner should be called upon to look into the matter and arrive at a decision.

Now we come to the fine art of issuing instructions for services as they fall due. Practices vary a great deal in this respect, depending on the paperwork that has been adopted. Furthermore, scheduling of work for several craft teams becomes fairly involved. A choice has to be made between attempting the control by a central planner or splitting this function into separate groups according to crafts. The former has the advantage that all the data is centralised but its operation requires duplicate paperwork. Another objection often raised is that the central planner is ineffective. It also often becomes a problem to ensure that completed work tickets return to the planner so that he can post the results. The final decision is to be made in consideration of such factors as the availability of copying equipment, the number of craft teams involved and the degree of interrelationship between crafts. If craft teams can operate independently, even though a certain amount of duplication may exist, this is certainly more easily managed. Four basic approaches can be defined as follows;

(*a*) *A centrally controlled procedure*. A service-planning clerk maintains schedules that have been set up, either in the form of planning charts or visible-card indexes. He issues copies of instructions when necessary. With the service schedule completed the instructions would be returned and recorded on the history cards.

(*b*) *A functionally decentralised procedure*. The lubrication team is in possession of planning cards and supporting service instructions. The team plans its own work and implements the services. Similarly an inspection or an electrical team follows its instructions. Thus control is exercised over types of services instead of plant. As service charts are completed at the end of their period, they are sent to a central recording clerk who posts the services on the equipment history card for monthly, quarterly, or semi-annual periods at a time and files the schedules for future reference. As mentioned earlier, there is no point in recording individual uneventful services on the history card since the schedules already contain these and it would lead to over-handling of cards which would soon be filled with repetitious reference to regular services.

(*c*) *An intermittent-reminder procedure*. In this case the clerk only issues a weekly or monthly list of reminders for groups of activities. The team then assembles the instruction sheets from sets of manuals and plans its own work within the time limits set. The service can be assumed to have been performed uneventfully except when the clerk is notified otherwise. In that case he may have to issue a work order for rectifying whatever has been found to be faulty. No recording would be required for the recurrent service except bringing the history up to date at the end of the year with a single entry: "Regular services carried out as prescribed."

(*d*) A hybrid combination of these approaches.

All these procedures presuppose that service crews are not required to do repairs, except for minor adjustments. In all cases where a defect has to be repaired, a request for it would be issued and the repair team would take over. If this were not done and the service crews were required to do the repairs, the schedule would invariably be delayed to the serious detriment of the system. Repair jobs arising out of inspections are to be treated like other jobs that have to be included in a work schedule. Urgent cases would be given consideration, while the rest can await their inclusion in a non-rush work plan. Alternatively, a repair could be delayed so as to take place when the next service is due. Despite a certain risk that such a delay entails, this procedure can be recommended as being economical in manpower. Let us now examine these procedures in greater detail.

(*a*) A centrally controlled system can be easily operated by a visible-card index with indicators (coloured riders), such as the visible-card systems shown in Fig. 12 (p. 52) and Fig. 47 (p. 114). Visible-card indexes where cards are staggered so as to expose a side or the bottom margin can have an annual scale printed in a variety of ways. Their choice is a matter of personal preference. The visible scale is intended to show the date on which the service is due. A perfect alignment of the cards is here essential so as to enable the schedule clerk to find all the markers located in one line along the time-scale. Moving the marker to the next date is done either when instructions are issued or upon completion. These card systems in themselves are beyond reproach. However, they violate the principle that *uneventful routine service should not be* recorded except on collective schedules. On the latter, for instance, a schedule with 40 lines and 52 weeks, that can serve to record 2,080 services, will usually take about 500 events. To post these individually on separate cards would take between ten to fifteen hours of clerical work. On a spread-out schedule sheet the same number of postings would probably take less than one hour.

Among the disadvantages of the centralised scheme is the profusion of papers and transcriptions that it involves. Another aspect here is the assumption that service teams work like robots, performing the work as it is issued by the clerk. This may have a bad effect on morale. Besides, there is no assurance that service teams will work like robots.

(*b*) The functionally decentralised procedure allows the crew a much higher degree of self-planning. Once they are assigned a function, *e.g.* lubrication or inspection, they are expected to perform it as per schedule. In contrast to the previous procedure they record on the schedules the completed jobs as they proceed (Fig. 44). Frequencies beyond one month should appear on yearly charts.

Since the workload can easily be established in terms of hours, the manning of these crews is not left to chance. Whether one person alone or a team is entrusted with a regular function he or they can be held responsible for its performance.

Incidentally, this approach can serve as the basis for an incentive scheme. A daily load expressed in time could either be in the form of a high-task 480 minutes or a target figure above 480, say 600 standard minutes. The performance of the allocated daily task supported by sampling evidence, justifies payment of the bonus.

It is apparent that the "functionally decentralised" procedure lightens the clerical load to a considerable extent. The job of planned services should be able to proceed without the necessity of reminders, notices or work orders, as the discovery of impending failures should lead to the issue of repair work tickets with the repairs being dealt with promptly. With the exception of being more detailed and exacting, the operation of a planned maintenance system can be compared to routine maintenance where the schedule sheet serves as a *reminder*, a *job order* and a *recording* document.

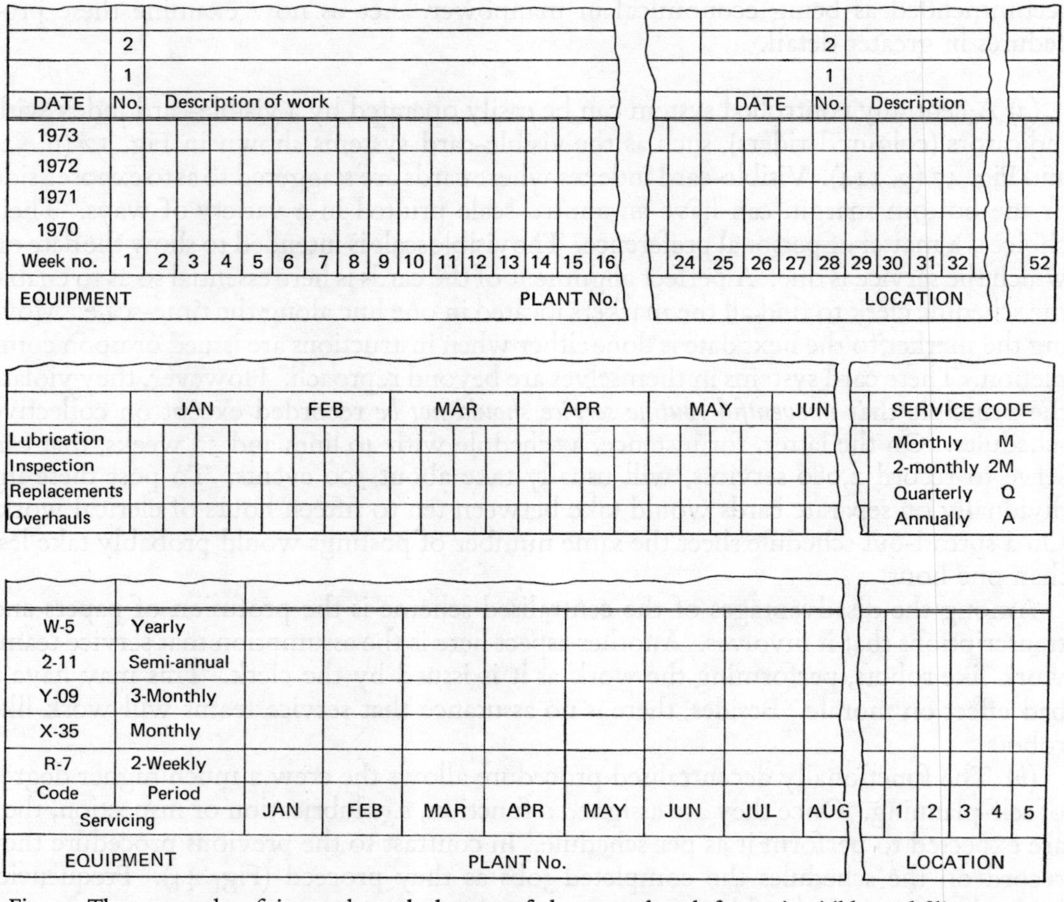

Fig. 47. Three examples of time-scales at the bottom of plant record cards for use in visible-card filing systems.

The advantage of this approach lies in the fact that required data for a number of units is kept in a concise form, often on one sheet only, for a whole year. Frequencies can be altered easily by making corrections on the instruction sheet. Varying workloads between teams, or seasonal variations can also be conveniently handled.

This procedure, as well as the other related ones, requires a good measure of follow-up. It is not enough to issue the instructions and to receive back a completed form duly signed. Supervisors must spot-check performance and the timing of jobs should be reviewed at intervals. These services are best performed when, for instance, an oiler is allocated jobs covering eight hours of work. In such instances the worker can set out on his service route with his whole day's work planned. The work sequence could either be set by the planning clerk or by the craft supervisor.

It is a useful practice to issue blank work-order forms to service teams and to require them to fill in the details as they discover repair jobs or imminent failures. These would be scrutinised by the foreman and would require his authorisation before being implemented.

Maintenance workers assigned to the central pool are here assumed to be interchangeable and anyone given clear instructions can follow them through.

(*c*) The third choice, the "intermittent-reminder" approach is sometimes employed. Service teams are issued with coded and numbered instruction sheets laminated in a protective sheath for repeated use. A list of services is prepared weekly by a clerk who issues it accompanied by the service instructions. The completed jobs are checked off on the list and at the end of each week the set is returned to the planning clerk. The results are regularly transcribed on to the respective equipment cards. Other details are common with either of the earlier mentioned procedures.

(*d*) A departure from these procedures is a combination of a schedule chart and a log book. The following example can best illustrate this approach:

> Every production department is allocated a certain amount of service time either weekly, monthly or quarterly.
>
> A schedule for the year is affixed to a wall-board at the entrance of each shop.
>
> When a serviceman is assigned to a certain production department, he goes there and consults the chart on the board, notes the machines due for servicing and proceeds to service them.
>
> On each machine a service record card is affixed in a suitable container-pouch, where lubrication charts and service records are also kept.
>
> Upon completion of each service, the maintenance man records it in the appropriate space, and puts the records back into the container-pouch. Before leaving the shop he records the services performed on the wall-chart, next to the entrance.

This approach is advocated in order to provide an on-the-spot service check in each department. Occasionally, machines carry a card pre-printed with the annual calendar (in days or weeks or months) and the serviceman is provided with a train-conductor type of perforator to punch the appropriate date. Thus, whenever a machine is serviced, the last date of service can be seen on the tag cards (*see* Fig. 48). In fact, these tags can also be pre-marked with the service dates for a whole year in advance.

As we can see, planned maintenance is more elaborate and detailed than routine servicing. Under this system the incidence of unforeseen failures is all but eliminated by concentrating on service-as-the-plant-requires. The preventive effect of planned servicing is here heightened by more specific instructions and by more frequent attention. Except for the definitive requirement of "prevention-at-any-cost," this system is almost fail-proof. To make it so, requires the preventive maintenance approach discussed below.

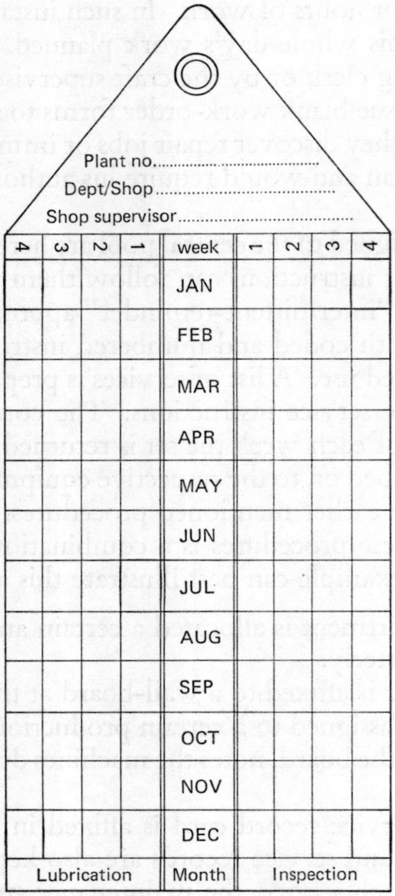

Fig. 48. Service control tag for attachment to equipment.

SCHEDULING OF PREVENTIVE MAINTENANCE

Preventive maintenance, notwithstanding the widespread misuse of this term, refers to a system of planned work which incorporates inspections and replacement instructions intended *to eliminate the occurrence of defects.* When considering essential installations and life-endangering equipment this service is to be implemented "at any cost," meaning that no effort should be spared to discover and eliminate any source of possible failure. While planned maintenance provides the normal service that is needed, preventive maintenance endeavours to discover developing faults or failing parts.

It is regrettable that, particularly in the U.S.A., simple routine schedules are called "preventive maintenance." This term should be reserved for use in cases where prevention of breakdowns is the *prime objective* of the service.

Procedures for the planning and scheduling of preventive maintenance are not very different from those relating to planned maintenance. There are however some built-in features which would make it almost "infallible" in achieving trouble-free operation.

The main difference in actual implementation lies in the inclusion of detailed *preventive inspections* which have to be carried out between the planned service dates. This system requires a good "memory core" to initiate these inspections. The "tickler file" is such an operating gadget. Let us see how it works.

If we were to take a diary for the current year and mark in advance all services due on the respective dates and pages, we could gradually leaf through it and discover as we progress what work is due on a certain day or within the current week. The tickler file is the same idea except for the fact that the pages are in the form of cards, and that *each machine* has its *set of cards*. All the cards carrying the instructions and recording of the work to be done are placed in a box file divided into the twelve months of the year (*see* Fig. 49a).

The cards are not fixed in a certain sequence within the set. When a certain job has been performed and recorded on the card, it is moved back within the set to a spot representing the date of the next service. Thus cards that have reached the front of the pack are dealt with first and then individually removed to the back according to the length of the given period. There are dividers in the box file to indicate the first,

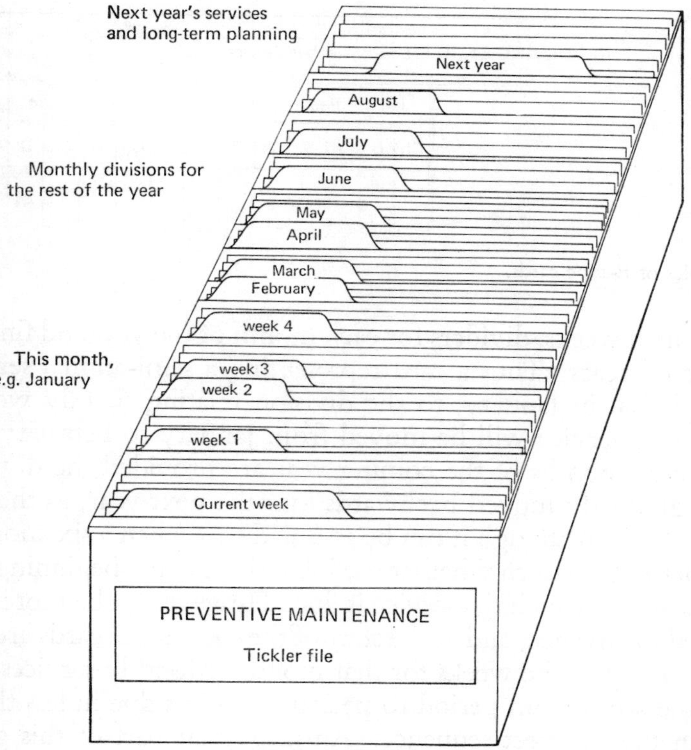

Fig. 49a. The revolving card file, otherwise known as a "tickler file."

Annual	REVOLVING CARD FILE—Preventive Service							Plant no.
INSTRUCTIONS	1970		1971		1972		1973	
	Ch	R	Ch	R	Ch	R	Ch	R

Semi-annual	REVOLVING CARD FILE—Preventive Service							Plant no.
INSTRUCTIONS	1970				1971			
	Jan		Jul		Jan		Jul	
	Ch	R	Ch	R	Ch	R	Ch	R

Monthly	REVOLVING CARD FILE—Preventive Service															Plant no.
INSTRUCTIONS	Jan		Feb		Mar		Apr		May		Jun		Jul		Aug	
	Ch	R	Ch	R	Ch	R	Ch	R	Ch	R	Ch	R	Ch	R	Ch	R

Fig. 49b. Some examples of B-type cards.

second, third and fourth weeks, dividers for each month of the year and finally dividers for the next couple of years. Thus a card representing a semi-annual service will be moved, after completion in January, to the division relating to July work. A card representing a monthly service will be moved from January to February. If no sub-dividers for the twelve months of the coming year are provided the dividers for the current month are gradually moved backwards into the next year, as the services are progressively completed. Although it can be assumed that when a six-monthly service is due, it will not matter too much whether the job is done at the beginning, the middle or the end of the month in which the service is due. However, to be more accurate the next date is recorded in advance, and at the start of the month the cards are arranged in a sequence corresponding to the weeks for that month. Monthly services will thus be kept within the same week from period to period. Services due in less than a month will also be kept in their correct sequence. An important part of this gadget is the proper design of the forms (Fig. 49b).

The following points need to be emphasised:

(*a*) Preventive maintenance should be based on facts and statistically proven frequencies. (This point again diverges from the popular notion of preventive maintenance.) It would be wrong and costly to replace belts, light-bulbs, filters, etc. on the basis of hunches. If preventive maintenance is to be economically viable, statistical data of experiments and/or past performance must be analysed and conclusions drawn accordingly.

(*b*) The effectiveness of the system can only be as good as the people who run it, and therefore no effort should be spared to train all personnel concerned. Plant operators should report when sounds, sights or smells seem indicative of impending trouble. Repair men should keep alert for signs of possible weaknesses.

(*c*) Preventive maintenance should be applied selectively. For instance, it is not the right approach where there are a hundred identical units in operation, *e.g.* looms, sewing-machines, etc., and some of these not even fully utilised. Preventive maintenance, however, is vital for boilers that provide steam for a whole factory and for furnaces or kilns that take two weeks to cool and one week to put back into operation and it is equally vital on production lines involving dozens of interlinked machines.

(*d*) A system can be called "preventive maintenance" only if all possible steps are taken to ensure continuous, trouble-free operation. These steps would include a very thorough review of the equipment with the intent to discover possible points of failure and the means of eliminating them. Defects can be discovered while they are developing and, where possible, prediction of failures should be based on instrument measurements or statistical research.

To ensure trouble-free operation between planned services, clearly defined inspection checklists should be prepared and implemented. While the planned service is set until revised, inspections can take place at any odd intervals *in addition* to the defined service frequency. Inspections may be visual, operational or physical. The last group has greatly progressed in the last two decades and the potential of a vast range of sophisticated instruments should be exploited to its fullest extent.

Figures 50a and 50b show a slightly changed version of the normal British Standards acceptance charts. The present example allows recording of the measurements on the form itself within the frames of the diagram.

Machines whose accurate performance is critical to the company will have such inspections at regular intervals. Checklists can also be employed (*see* Fig. 51). These forms will be filed either by plant number sequence, or by machine groups or by departments.

In preventive maintenance, lubrication schedules and the instructions relating to them are also more detailed and specific. Lubrication diagrams such as shown in Fig. 52 are used. The surrounding frames indicate the prescribed frequencies and the circles and triangles the types of service. Figure 53 shows a one-monthly lubrication schedule. It contains five weeks so as to accommodate any configuration of dates during the year. Twelve such sheets will be issued in a year.

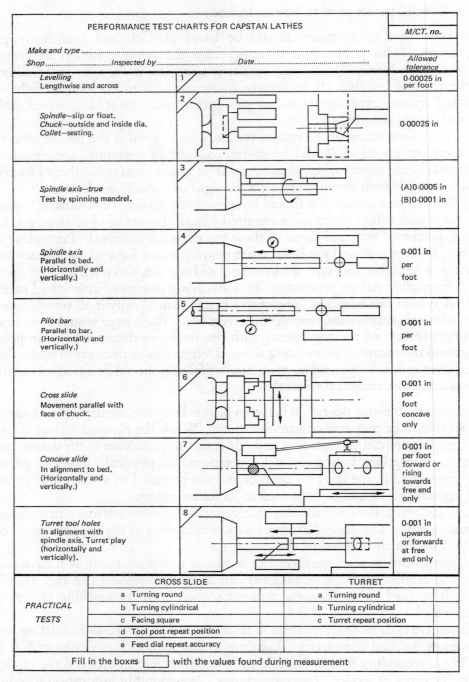

PERFORMANCE TEST CHARTS FOR CAPSTAN LATHES		*M/CT. no.*
Make and type ...... *Shop* ...... *Inspected by* ...... *Date*		*Allowed tolerance*
Levelling Lengthwise and across	1	0·00025 in per foot
Spindle—slip or float. *Chuck*—outside and inside dia. *Collet*—seating.	2 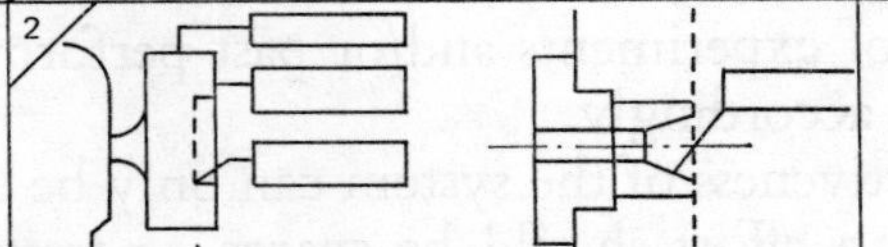	0·00025 in
Spindle axis—true Test by spinning mandrel.	3 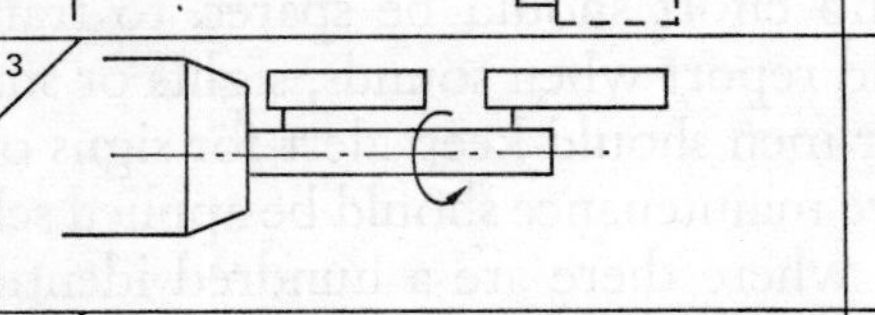	(A)0·0005 in (B)0·0001 in
Spindle axis Parallel to bed. (Horizontally and vertically.)	4 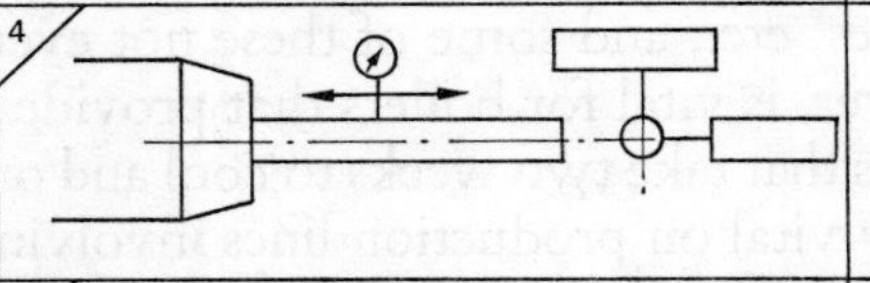	0·001 in per foot
Pilot bar Parallel to bar. (Horizontally and vertically.)	5 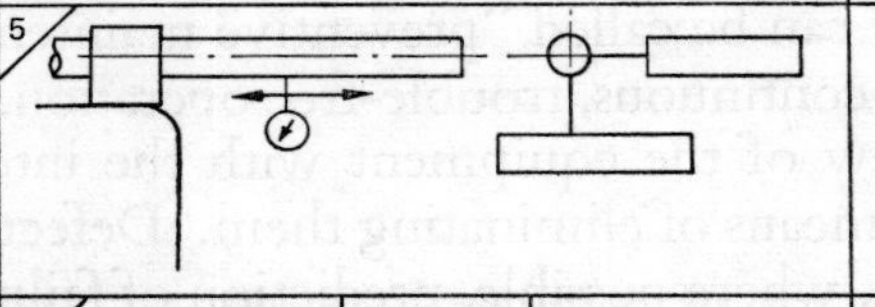	0·001 in per foot
Cross slide Movement parallel with face of chuck.	6 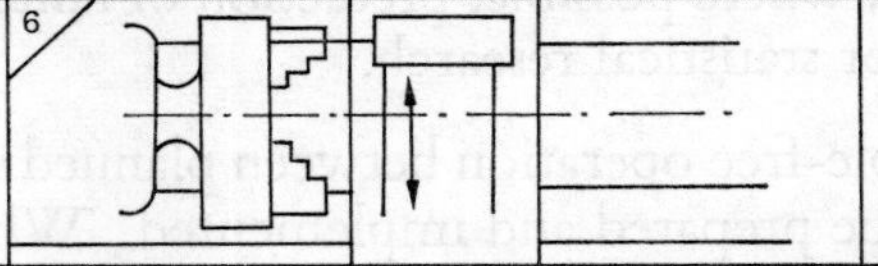	0·001 in per foot concave only
Concave slide In alignment to bed. (Horizontally and vertically.)	7 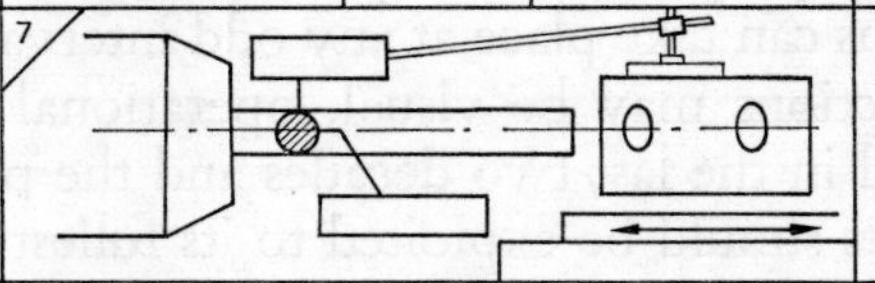	0·001 in per foot forward or rising towards free end only
Turret tool holes In alignment with spindle axis. Turret play (horizontally and vertically).	8 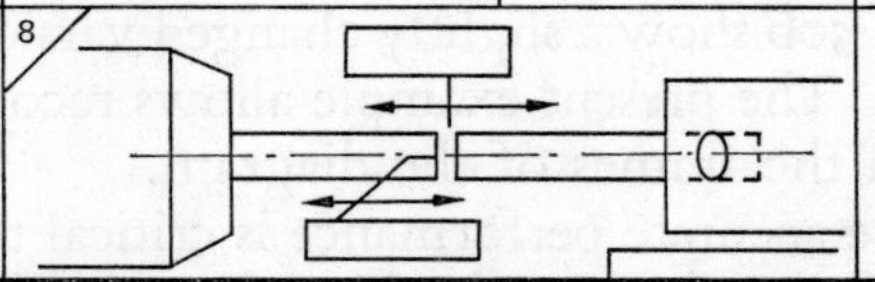	0·001 in upwards or forwards at free end only

	CROSS SLIDE		TURRET	
PRACTICAL TESTS	a Turning round		a Turning round	
	b Turning cylindrical		b Turning cylindrical	
	c Facing square		c Turret repeat position	
	d Tool post repeat position			
	e Feed dial repeat accuracy			

Fill in the boxes ☐ with the values found during measurement

Fig. 50a. Example of a performance test chart.

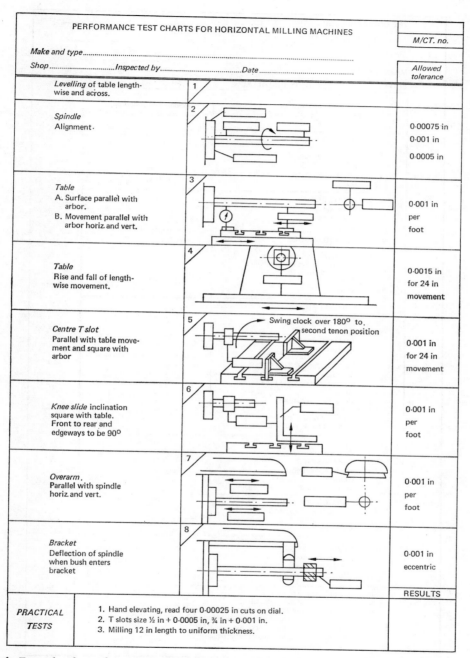

PERFORMANCE TEST CHARTS FOR HORIZONTAL MILLING MACHINES		M/CT. no.
Make and type............		
Shop............*Inspected by*............*Date*............		*Allowed tolerance*
Levelling of table length-wise and across.	1	
Spindle Alignment.	2 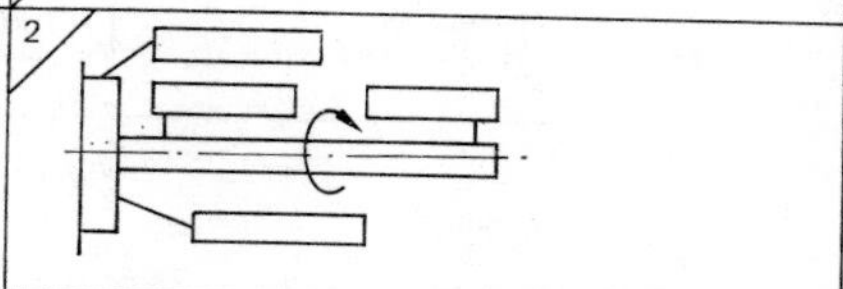	0·00075 in 0·001 in 0·0005 in
Table A. Surface parallel with arbor. B. Movement parallel with arbor horiz. and vert.	3 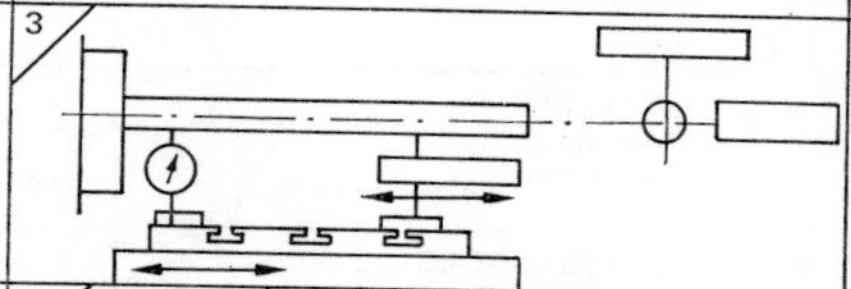	0·001 in per foot
Table Rise and fall of length-wise movement.	4 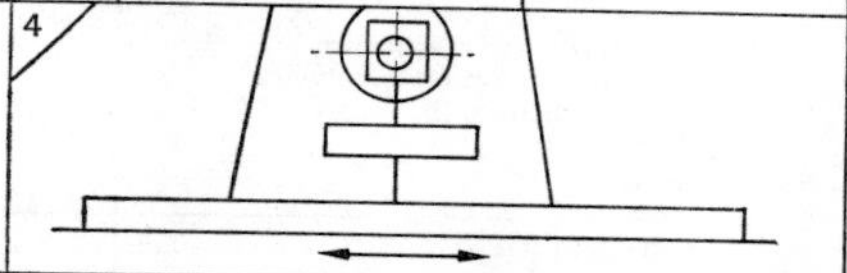	0·0015 in for 24 in movement
Centre T slot Parallel with table move-ment and square with arbor	5 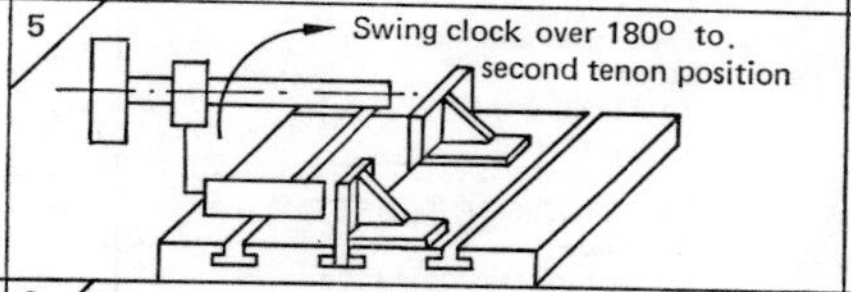	0·001 in for 24 in movement
Knee slide inclination square with table. Front to rear and edgeways to be 90°	6 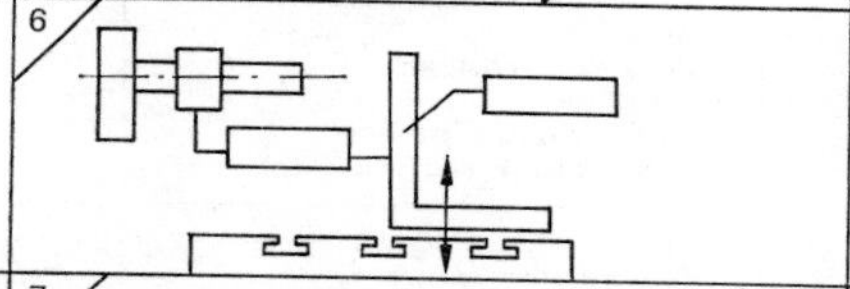	0·001 in per foot
Overarm. Parallel with spindle horiz. and vert.	7 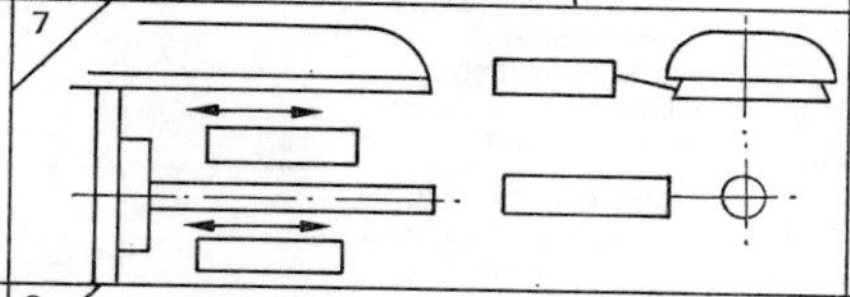	0·001 in per foot
Bracket Deflection of spindle when bush enters bracket	8 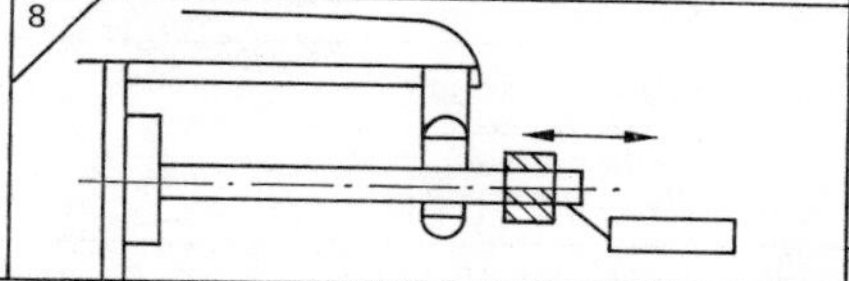	0·001 in eccentric
		RESULTS
PRACTICAL TESTS	1. Hand elevating, read four 0·00025 in cuts on dial. 2. T slots size ½ in + 0·0005 in, ¾ in + 0·001 in. 3. Milling 12 in length to uniform thickness.	

Fig. 50b. Example of a performance test chart.

THE UNIMATIC ENGINEERING CO. LTD.

VISUAL CHECK-UP SHEET

LATHES

M/C Tool No.

MAKE AND TYPE

SHOP DATE INSTALLED DATE

	Parts to be checked:— ✓—O.K.; X—Faulty; XX—Very bad	*Remarks on condition, lubrication, and operation*
	HEADSTOCK Bearings Gears and spindle Pilot bar Chuck (pneum, hand) or collet Guards	
	LATHE BED Slideways Bed covers Guards	
	SLIDING SADDLE Slides Lead screw and mechanism Feed mechanism (a) lengthwise Feed mechanism (b) across Stop carrier Covers	
	CROSS SLIDE Slides Feed screw and index dial Stops Tool post, front Tool post, rear	
	TAILSTOCK OR TURRET Base slide Capstan (or tailstock) slide Turret Stops and mechanism Star wheel Feed mechanism, traverse Tailstock barrel and lock (C.L.)	
	DRIVE Forward, stop, reverse Speed handles Brakes and clutch Gearbox Belts and guard	
	LUBRICANTS Oil pump and indicators Coolant pump Suds pipes	
	MISCELLANEOUS Lighting fixtures Electric switchgear Electric motor	

(a) What spares are required?
(b) Is a new coat of paint advisable?
(c) Is there excessive vibration when on heavy load?
Is there excessive vibration when on light load?
(d) Overall condition: Very good, Average, Poor.
(e) This plant is used: Continuously, Frequently, Occasionally.

Fig. 51. A visual checklist for inspecting lathes: a typical type. A form that can easily be adapted for repeated use by extending the "remarks" column to the right.

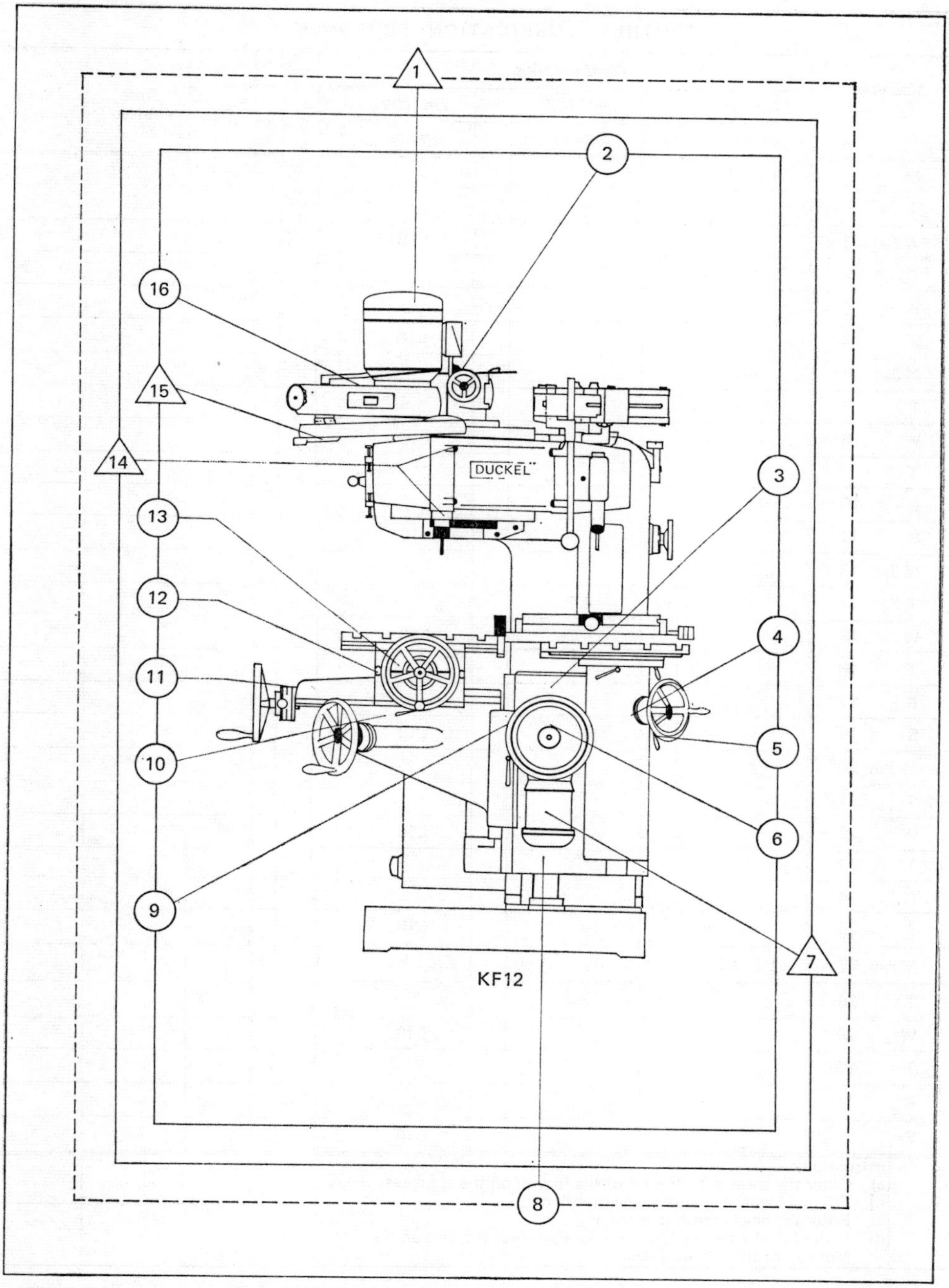

Fig. 52. A lubrication diagram in which the frames refer to different periods of service and the circles and triangles to the types of service.

MONTHLY LUBRICATION SCHEDULE										
	Machines	*Plant unit nos.*								
		No. 101		*No. 102*		*No. 103*		*Sig.*	*Time allowed*	*Supv. check*
Date		*Service*		*Service*		*Service*				
WEEK 1	M	L3 A5								
	T			A4						
	W Jan. 1					L16				
	T 2									
	F 3									
	S 4					L16				
WEEK 2	M Jan. 6	A5								
	T 7									
	W 8					L16				
	T 9									
	F 10									
	S 11					L16				
WEEK 3	M Jan. 13	L3 A5								
	T 14			A4						
	W 15					L16				
	T 16									
	F 17									
	S 18					L16				
WEEK 4	M Jan. 20	A5								
	T 21									
	W 22					L16				
	T 23									
	F 24									
	S 25					L16				
WEEK 5	M Jan. 27	L3 A5								
	T 28			A4						
	W 29					L16				
	T 30									
	F 31									
	S					L16				

Instructions:-
(a) Enter the sheet with the 1st of the month on the appropriate day
(b) Fill in all working days of the month and
(c) Enter the next month in week 1
(d) If the 1st of a month falls on a Sunday enter the 2nd on the Monday of the second week.

Fig. 53. A monthly lubrication schedule incorporating five weekly periods to accommodate any configuration of dates.

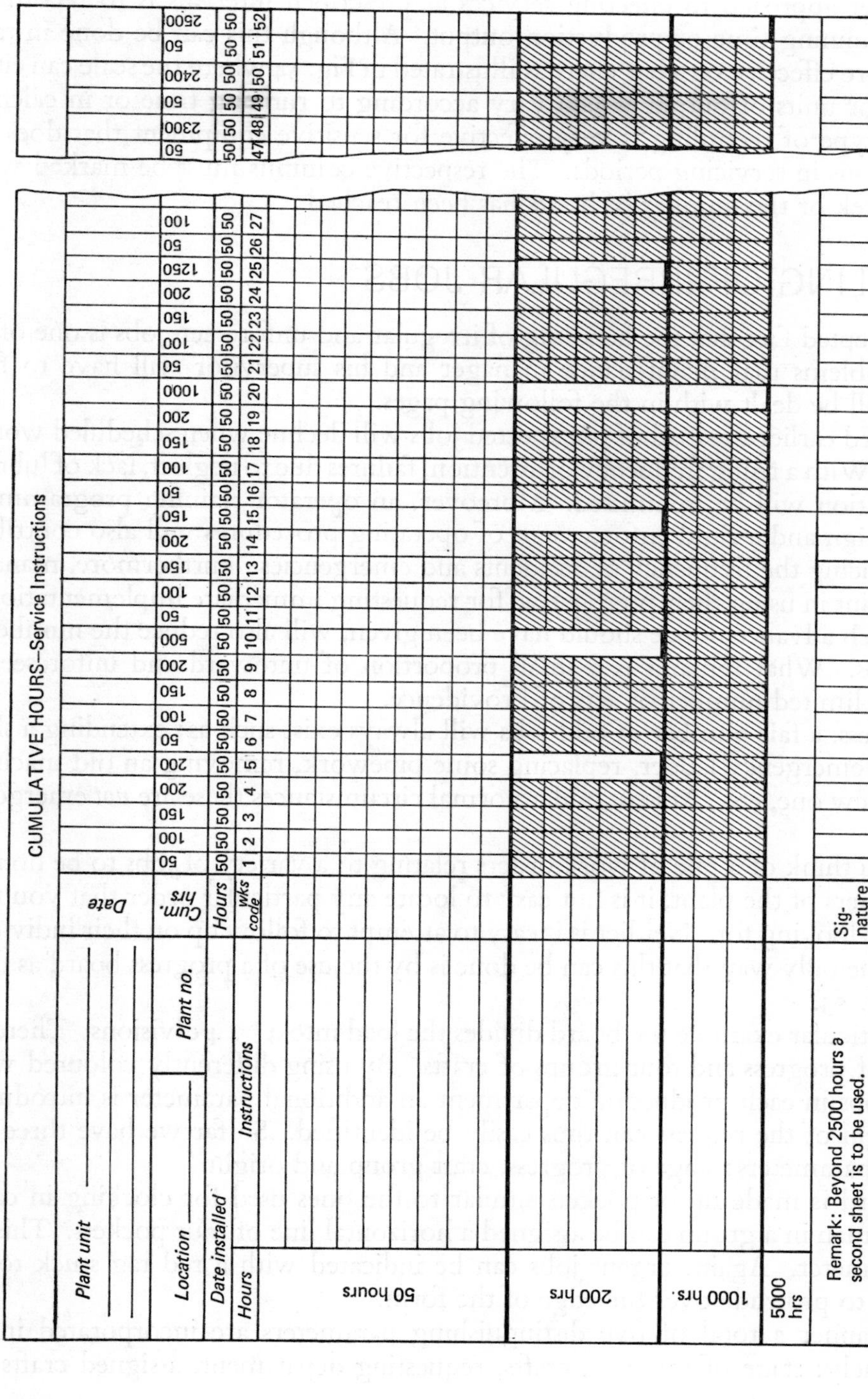

Fig. 54. Service instruction sheet used according to cumulative hours of running time. The respective columns must be marked when either the week or the number of hours has been reached.

Yet another approach to effecting services at prescribed intervals is to record the cumulative running time or production output. Although this can be done in table form it is more effective on a bar chart as illustrated in Fig. 54 where the scale can either be in hours or units of output which vary according to running time or in calendar dates. This type of recording is very effective for sensitive equipment that does not allow variations in servicing periods. The respective columns must be marked when either the week or the number of hours has been reached.

SCHEDULING OF IRREGULAR JOBS

It is an accepted fact that the handling of irregular and unforeseen jobs is one of the toughest problems that maintenance manager and his supervisor will have to face. This issue will be dealt with in the following pages.

As described earlier the load of unexpected jobs will decline when scheduled work is carried out. With a rising frequency of attention, failures due to neglect, lack of lubrication or attention will be minimised. Moreover, an operator training programme, a safety campaign and a rigid enforcement of operating procedures will also contribute towards reducing the frequency of accidents and emergencies. Furthermore, management's restraint in using their prerogative for requesting immediate implementation of jobs, on which advance notice should have been given, will also reduce the number of irregular jobs. What remains is a small proportion of untoward and unforeseeable events and a limited number of acts of Providence.

Nevertheless, a fair number of odd jobs will always exist such as: extending a shed, installing an emergency ladder, replacing some pipework, removing an old machine, installing a new one, etc. At least under normal circumstances these are *not* emergency jobs.

When you think of a pile of work orders relating to a variety of jobs to be done in the four corners of the plant, it is not easy to locate any particular order that you may happen to be looking for. Neither is it easy to attempt to follow up on their individual progress. The only way that this can be done is by the use of a progress board as illustrated in Fig. 55.

In this particular example the board divides the load into 4 × 4 divisions. There are four stages of progress and four groups of crafts. By using differently coloured work orders for jobs in each production department an additional parameter is introduced, and the origin of the request can thus easily be identified. So far we have three distinguishing parameters: stage of progress, craft group and origin.

If the board is made up of pockets similar to the ones used for clocking-in cards, every tradesman in a group can be assigned a horizontal line of four pockets. This is a further parameter. Again, urgent jobs can be indicated with a red tag stuck to the corner so as to protrude over the edge of the form.

In this manner a total of five distinguishing parameters are incorporated in the system, namely: stage of progress, crafts, requesting department, assigned craftsman and urgency.

In this form the board can serve as a job-allocation device. But a great advantage is that in cases where there is an enquiry concerning the progress of a job, any job card can easily be located since the position of every job can easily be followed on the board.

In large departments each craft group will have a separate progress board, preferably served by a despatcher. Incoming jobs will be kept for the supervisor's attention until he clears them by initialling them or by recording on them his special instructions. The job is then immediately allocated to one of the tradesmen by placing the work order in

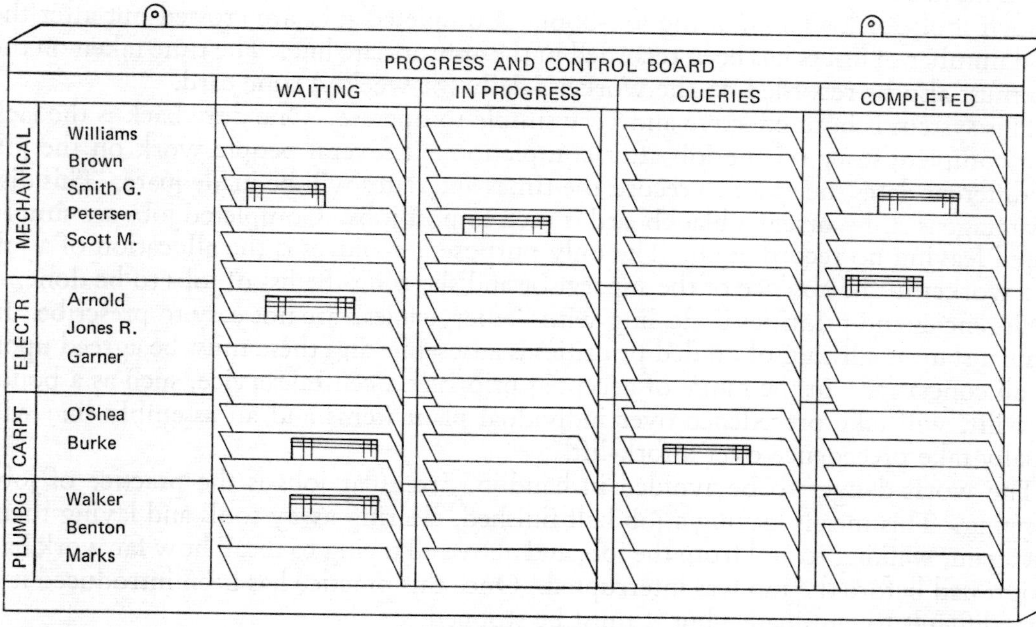

Fig. 55. Progress and control board for work orders showing four stages of progress and four groups of crafts.

his respective slot. The supervisor is therefore not tied to his desk and the board will act for him in his absence. Completed tickets are kept on the board until the details of the job are recorded on them and a brief description is transferred to the respective plant history cards.

Work orders which cannot immediately be carried out either because parts are not available or instructions are not clear or the machine cannot be shut down for servicing are placed in the "queries" column. If this is not done there is a danger that they will forever be in the "waiting" box. This technique has proven itself easily operable and very effective.

The board may take the form as here illustrated or instead of pockets it may simply have spring-clips attached to a board divided into separate fields. Some boards are home made by slotting four strips of wood diagonally and into these the orders are inserted.

At least two additional planning devices must be mentioned here, namely the diary log-book and the blackboard-posting procedures.

It is often claimed that since many jobs are requested by telephone or verbally, it would be wrong to rely entirely on devices using only individual written requests. The open diary is then suggested. An open notebook is kept on the supervisor's desk and all incoming calls are recorded therein. The supervisor or the clerk receiving a telephone call records the details and the supervisor assigns the job to one of his men whose name is then recorded next to the job. When this worker becomes available he is informed about his next job as derived from the log-book. In case nobody is at the desk, he can himself look for his name in the log-book. Completed jobs are crossed out after the total number of hours has been recorded in the appropriate line. The time taken on the job must also be recorded on the worker's daily (or weekly) time card.

This system is both effective and very simple to operate. One drawback is the lack of a complete story on the job after completion. If several people work on the job on different dates the space to record the times and dates will be inadequate. This disadvantage is still worse if a blackboard is used to post jobs. Completed jobs are simply erased leaving no record at all. The only purpose it achieves is the allocation of a job to a worker in the absence of the supervisor and showing the list of jobs to be done.

Urgencies and priorities in dealing with "emergencies" are not easy to prescribe. In some instances a listing of graded priorities is advisable, and these must be agreed upon by all concerned. In the range of accepted priorities a central service, such as a boiler or a lift, will take precedence over individual plant items and an assembly line will usually take precedence over a fork-lift.

The worst danger to be avoided in handling irregular jobs is the practice of job-hopping. This means leaving a job half finished, clearing away tools and laying them out again, walking to and from the job, and above all trying to recall how far work had progressed before the job was interrupted. Once this practice has been introduced it is very difficult to eliminate—but it must be stopped.

Irregular jobs should be pre-planned as much as possible. Parts, materials, supplies, special tools, etc. should be prepared in advance. Some companies use preparation bays where all requisites are deposited. Only then does work start. The ordering of parts or any attempts to purchase parts locally should be done well in advance. Old hands will counter this argument by saying that often it is not known what is needed until the job is half-way through. This is true, but these are not the cases we are handling most commonly.

To counter this latter argument there is at least one measure we can initiate, namely the pre-planning and issuing of replacement kits which would anticipate demand. A case in point is the overhaul of fork-lifts or tractors. To minimise walking time to stores, complete kits on wheeled stands are prepared and placed next to the repair bay. During work all that is needed is taken from the kit. Upon completion the remainder is taken to the stores and the kit is fully stocked up again.

Another feature of irregular jobs is the requirement for different crafts to converge on a certain spot at certain periods, thereby making them interdependent. This poses problems of co-ordination which no textbook can solve. There are, naturally, procedures that alleviate the problem such as pre-printed duplicates of orders and instant communication channels which enable suitable dates to be set by common consent.

The use of Gantt charts for work planning is not limited to irregular jobs, but it appears to have more relevance for shutdown planning, project work and jobs that occur repeatedly and must be carried out in a standardised sequence (*see* Figs. 56a and 56b). In this instance a recurring locomotive inspection requires the co-ordination of a large staff within a short period of time. The simplest way to show their interdependence is by a Gantt chart.

A good case can be made for using charts, A4-sheet size, for a craft group. In most cases a weekly chart for a team of eight to twelve people will be found to be practicable. However, experience shows this procedure never survives for prolonged periods. Drafting the charts and keeping them constantly up to date is a chore and so they are soon discarded. A computerised daily or weekly print-out is an excellent substitute. In factories where production is closely programmed and controlled by a computer, applying a similar procedure to maintenance is simply a by-product.

A computerised system of work control affords many advantages. Regular and routine jobs get pre-printed job cards and master cards for regular services are also pre-printed. Individual services with pre-coded frequencies can also be printed by computer. Last but not least, daily, weekly and monthly print-outs can be obtained for work done or for work to be done.

Electronic Data Processing (EDP) where available and economical is both practical and advisable. Pre-printed punch cards can be marked with magnetic-sensitive pencils and programmed in proper sequence on a daily or weekly print-out. Obviously the system requires code numbering for most details and time estimates that the computer can summarise. For a start this will raise some problems. But once such a system is introduced many advantages accrue. Statistical analysis by crafts, cost centres, departments and tradesmen will present no problems. The different summaries of hours and costs can be easily done by processing the same coded data through the tabulator in different sequences.

The application of this system is feasible when the company is already operating an EDP system, either by having its own installation, by renting time, or by using commercial data-processing centres. Another condition to be met is the size of the maintenance team. Undoubtedly, for a total group of over fifty workers an EDP system is practicable. This would extend over the stores activities as outlined in Chapter 7, p. 137–150.

Those who are adversely inclined to the use of a computer should realise the tremendous advantages to be gained. Every coded and recorded item can be used, classified and reclassified effortlessly. The information is easily retrievable and the print-outs are conveniently arranged in ascending or descending order of job numbers and dates, by departments, craft sections or workers.

The only snag, and one that irks many computer-users, is the fact that reports are often long delayed and frequently overdue. It is suggested that the cut-off date for

DIESEL–ELECTRIC LOCOMOTIVE 150 HOURS' SERVICE					
Job no.	*Work description*	*Time (min.)*	*Craft*	*Crew*	*Man minutes*
	Mechanical				
A	Check valve gears and pump racks	25	Mech.	2	50
B	Remove oil filters and strainer bowls element. Replace reconditioned ones	45	Mech.	3	135
C	Remove clean and replace governor strainer	10	Mech.	1	10
D	Top up compressor, clean oil bath, check drive bolts	45	Mech.	1	45
E	Remove air mazes, clean ducts, replace clean mazes	45	Lab.	2	90
F	Clean locomotive	75	Lab.	2	150
G	Top up suspension	30	Help.	1	30
H	Check top up axle boxes	30	Help.	1	30
T	Testing	30	Mech.	1	30
	Electrical				
I	Check batteries Check main generator brushes	30	Elect.	1	30
J	Check blower motor brushes Check cubicle	120	Elect.	1	120
K	Check traction motor brushes	90	Elect.	1	90
	Air brake				
LM	Change brake shoes, adjust brake	60	Fitter	2	120
N	Check automatic and independent brake valves	90	Fitter	1	90
					1020
	Remarks This inspection period is repeated approx. once every seven or eight days, approx. forty times per year. Total man-hours to be allocated is 40 x 17 hrs = 680 man–hours per year				=17 hrs

Fig. 56a. Team schedule for 150-hour locomotive inspection, requiring the co-ordination of a large number of staff for a short period of time.

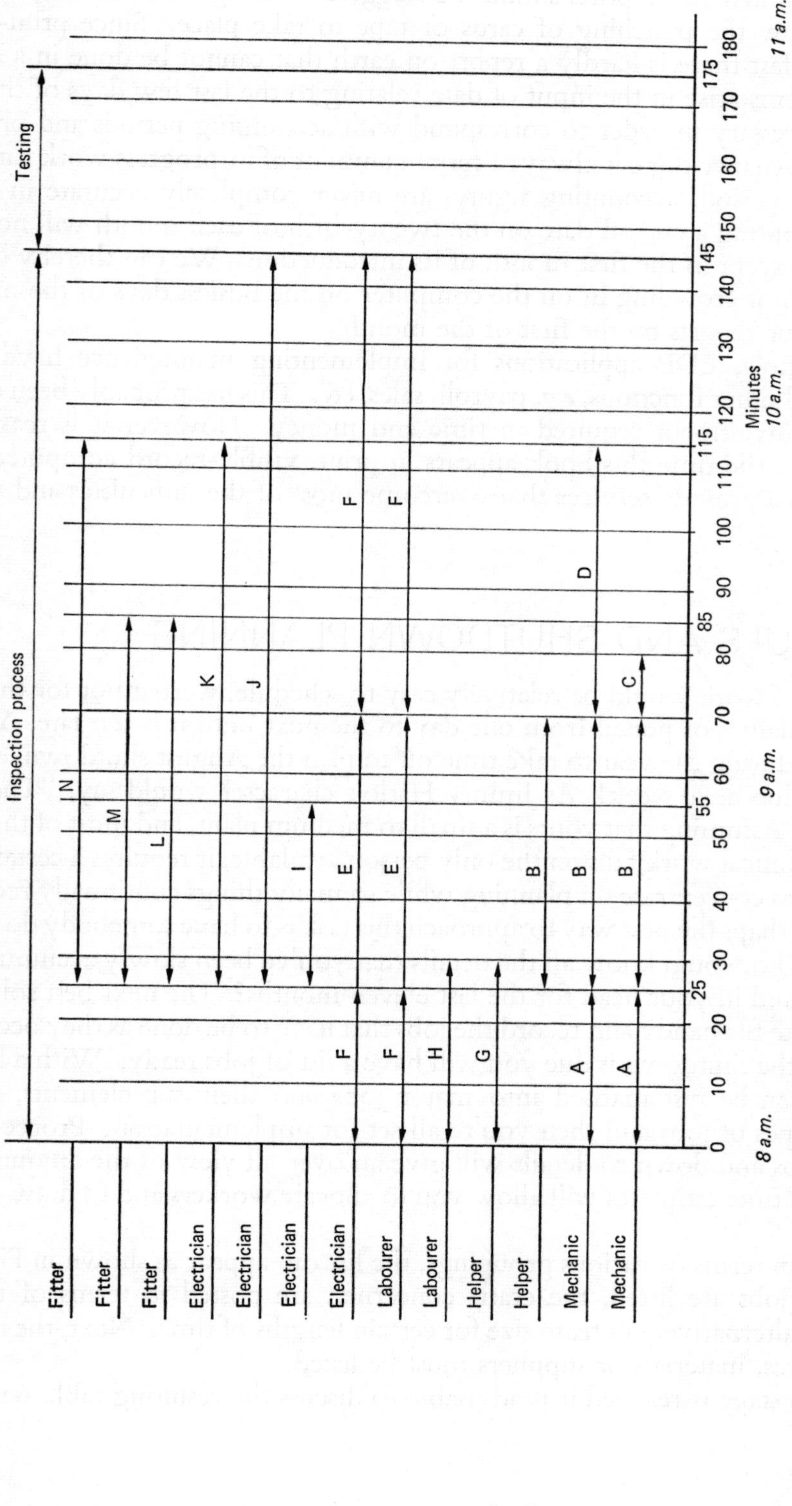

Fig. 56b. Job sequence for locomotive inspection; team activity chart.

processing maintenance reports should be staggered through the last five days of the month to allow the punching of cards or tape to take place. Since print-out is so astoundingly fast there is hardly a report on earth that cannot be done in a couple of hours. Problems arise in the input of data relating to the last few days of the month, which are necessary in order to correspond with accounting periods and procedures. Since in maintenance there is always a certain amount of in-progress work carried over into the next period, accounting figures are never completely accurate in any case. Therefore adopting a cut-off date on the twenty-fifth of each month will not make it less accurate except in the first month of its introduction. We can thereby obtain the advantage of not crowding in on the computer on the busiest days of the month and thus obtain our reports by the first of the month.

On the whole, EDP applications for implementing maintenance have seriously lagged behind other functions, *e.g.* payroll, sales, etc. This has probably been due to the considerable investment required in time and money. However it is reasonable to expect that by the time this book appears in print, visible-record computers (VRCs) may be able to provide services that overcome most of the difficulties and objections raised earlier.

OVERHAULS AND SHUTDOWN PLANNING

This type of work would be relatively easy to schedule, were it not for the fact that doing it is usually postponed from one day to the next until it is too late. You're too hard pressed during the year to take time off to plan the August shutdown, and lo and behold! it's due next week! As Jimmy Hatlo's character would say, "They'll do it every time." Assuming that yours is a small to medium plant, and most of the management and technical work falls on the only person available, it requires a certain amount of discipline to concentrate on planning while so many things continually require your attention. Perhaps the best way to approach this task is to have somebody do it for you, but who? Who would know all the details that you've been slowly accumulating and carrying around in your head for the last eleven months? The next best solution is to keep a separate file handy and record the jobs that have to be done as they occur to you. By the time the shutdown is due you will have a list of jobs ready. With a little effort a crude list can be systematised into major jobs and their sub-elements, groups of machines, types of jobs and then you're all set for implementation. Proceeding from the larger jobs and down to details will give an over-all view of the amount of work involved and time estimates will allow you to allocate workers and to draw up a Gantt chart.

Thinking in terms of various groupings, the list can appear as shown in Fig. 57.

Once the jobs are listed, the crafts concerned are posted in terms of man-hours required, or alternatively in team size for certain lengths of time. Next, the main components, spares, materials or suppliers must be listed.

When this stage is reached it is advisable to discuss the resulting table with all craft foremen involved. This is sure to yield additional ideas and will most certainly elicit useful comments.

The next hurdle is to ensure that everything needed is available in the stores when the time comes. This preparatory work must be done *well in advance*. Failure to prepare the necessary items in advance is the greatest obstacle to effective project work.

This procedure can be formalised in large plants, where the sequence really starts with having an appropriation allocated at budget time for all major foreseeable projects or in some cases seeking approval for these during the year. If a budget estimate is required the request must be preceded by considerable preparatory work consisting of labour and materials estimates. The trouble with these estimates is that the form in

LIST OF MAJOR AND OVERHAUL JOBS DURING SHUTDOWN — For the year..................

Loctn.	Major jobs and their sub-elements	Craft workload expected hours						Parts/tools materials/ required
		Mech.	Electr.	Builders	Carpen.	Painter	Total	
Department A								

Fig. 57. A list used to record jobs to be carried out during the annual shutdown.

which budgetary requests are set out is not identical to what is needed for scheduling or job-preparation purposes and so the listing has to be done again. It appears to be simple common sense that you cannot schedule on what, in fact, is a shopping list and vice versa. It is nevertheless suggested that we perform this part of the preparation efficiently enough to save ourselves the same trouble next time—for as sure as night follows day there is going to be a next time.

The actual scheduling is really not difficult once the listing of jobs is clear and detailed enough. On the principle of the Gantt chart a simple diagram along a time scale can usually be drawn (*see* Fig. 58) provided that time estimates have already been made.

Depending on the expected length of jobs or the total period of a shutdown, the scale could represent days, half-days or quarter-days. Or, the scale could represent twenty-four-hour periods duly subdivided into shifts or hours if work round the clock is anticipated.

The line drawn into the chart could be either single, double (in two parallel lines) or treble, representing the size of the team. If the specified job is "pump overhaul," the line could be divided by symbols and the stages of the job can be marked, such as: dismantle, recondition, reassemble and test. Since this sequence may involve different sizes of crews the suggested single or multiple line can easily indicate this.

A more complicated part of this schedule is the intercraft co-ordination. Again, this chart can assist us in this too. In fact, barring PERT–CPM techniques, a Gantt chart is the simplest device that can ensure against serious oversights. Since it is an effective device for communication it can easily be transmitted and verified by all concerned.

An adequate number of copies can serve as:

(*a*) advance notices to all concerned;
(*b*) an appendix to an appropriation request;
(*c*) a notice to the stores;
(*d*) a complement to the work order;
(*e*) a list of job cards to be issued or a progress chart to be marked as work is completed;
(*f*) a reference document for repeat jobs.

The simplicity of this procedure makes it suitable for most enterprises. It may serve

GANTT CHART SCHEDULE

Description	Total man hrs	Crew	Mon		Tue		Wed		Thu		Fri		Sat			
			a.m.	p.m.	a.m.	p.m.	a.m.	p.m.	a.m.	p.m.	a.m.	p.m.	a.m.	p.m.	a.m.	p.m.

or, an alternative time scale can be used:

Monday			Tuesday			Wednesday		
a.m. shft	p.m.	night	a.m.	p.m.	night	a.m.	p.m.	night

Fig. 58. Examples of Gantt-chart scales applicable to maintenance.

as an excellent assignment for a junior engineer or an understudy to the maintenance supervisor or for the training of a future foreman.

In some instances it is necessary to follow-up on the servicing of a large number of identical units. A steady rate of implementation is then desirable since we can easily allocate part-time work according to availability, *e.g.* two to four hours a day. To keep track of the work a Progressing Table such as is shown in Fig. 59 can be used. The time-scale can be in weeks or a daily division may be used. In this approach the form will serve both as a schedule and a record of work done. The form can easily be made to accommodate distinct phases of work which are identical for all units. This form can serve equally well for sewing-machines, switchgear, lamp-posts or pumps.

A special chart can be developed for planning the maintenance of major continuous networks or installations that can be defined in a measurable quantity, for instance, in miles of roads, pipelines, mining conveyors, power networks, etc. We can adapt such a device for the scheduling of the maintenance of railway tracks. The horizontal scale will represent the track which has to be identified by sections. Assuming a gang of workers to be assigned to each section, vertical columns of the chart will indicate parts of their annual schedule, as shown in Fig. 60.

PROGRESSING TABLE													
Plant group			*Progress*	*w/e 5 Jan*	*w/e 12 Jan*	*w/e. 19 Jan*	*w/e 25 Jan*	*w/e 2 Feb*	*w/e*	*w/e*	*w/e*	*w/e*	*w/e*
Roller conveyor no. (Bauxite Comp R)			Planned units	15	15	15	12	15					
			Cumulative plan	15	30	45	57	72					
			Actual completed	12	14	15	12	11					
No. of units	Maintenance time/unit	Total hrs allocated	Cumulative compl.	12	26	41	53	64					
120	72·0 min 1·2 hrs	144	Variance (+) (-)	-3	-4	-4	-4	-8					
			Man hrs. delay	3·6	4·8	4·8	4·8	9·6	etc.				

Spinning frames (Dept. A, Anytextile Ltd.)			*Weekly progress*	*Strip for cleaning*						*Servicing*						*Reassemble*				
				w/e	w/e	w/e	w/e	w/e	w/e											
			Planned																	
			Cum. planned																	
No. of units	Maintenance time/unit	Total hrs allocated	Actual compl.																	
			Cum. compl.																	
	.............hrs		Variance (+) (-)																	
	.............min		Man hrs. delay																	

Fig. 59. Two alternative charts for following weekly progress of work where many units undergo similar treatment. These forms may be used both as schedules and as records of work done.

RAILWAY–TRACK MAINTENANCE SCHEDULE							
Location	*Mileposts along the track* 0 miles … 12 … 21		… 30 … 35 miles … 48			*Total man-weeks allocated*	
	Section A		*Section B*				
Months	*Operations*	*allocated man-days*	*Operations*	*allocated man-days*		*Operations*	*man-wks*
January	Track alignment Levelling Re-gauging Sleepers, ballast Hedges, ditches						
	Total		Total				
February	Level crossings Switches, sidings, etc. Crossings						
	Total		Total				
March							
						Total annual budget	
TOTALS for TRACK SECTIONS							

Fig. 60. A simplified version of railway track maintenance schedule. The track mileposts serve as the scale of the chart.

This idea can also be adapted for showing progressive overhaul of buildings or heavy installations. For instance, the plant installation of a chemical process may have to be repainted. This may involve stripping off the old paint, application of a protective coating and final painting. If this is to be done over a period of several weeks we can schedule it in advance as shown in Fig. 61. By spreading out the work plan in this form we can avoid the mental gymnastics that are otherwise required.

PROGRESSIVE OVERHAUL SCHEDULE							Year................................	
Location / *Periods Week-ending*	*Main divisions of plant*						*Total man-days*	
	Department A		*Shop B*		*Process C*			
	Operation	*Man-hours*	*Operation*	*Man-hours*	*Operation*	*Man-hours*	*Planned*	*Actual*
Date	Repair defects	Plan		Plan		Plan		
		Act		Act		Act		
Date	Scrape off old paint	Plan	Repair defects	Plan		Plan		
		Act		Act		Act		
Date	Base coating	Plan	Scrape off old paint	Plan	Repair defects	Plan		
		Act		Act		Act		
Date	Painting	Plan	Base coating	Plan	Scrape off old paint	Plan		
		Act		Act		Act		
Date	Colour markings	Plan	Painting	Plan	Base coating	Plan		
		Act		Act		Act		
Date		Plan	Colour markings	Plan	Painting	Plan		
		Act		Act		Act		
Date		Plan		Plan	Colour markings	Plan		
		Act		Act		Act		
Date	Testing	Plan	Testing	Plan	Testing	Plan		
		Act		Act		Act		
Date		Plan		Plan		Plan		
		Act		Act		Act		

Fig. 61. An example of a schedule for the progressive overhaul of plant buildings. Copies of such schedules should be distributed to the people concerned, such as area supervisors and craft groups.

This section would not be complete without reference to methods designated as network analysis, such as PERT, and critical path analysis. This technique has been made even more attractive by the use of modestly sized (and priced) computers. A large number of package programs are available from computer manufacturers and data-processing bureaux. This solution, of course, obviates the need to own a computer. Since the subject has been adequately covered in other treatises it would serve no useful purpose to duplicate those efforts.

Chapter 7

Maintenance stores

The provision of supplies to ensure the smooth performance of maintenance is a subject that presents a wide range of problems.

If we declare that maintenance is treated as a stepchild by some managements, it can equally be said that stores and suppliers are the stepchildren of many maintenance managers. It appears that some departments are able to manage somehow with their stores in constant chaos, with poor lighting, sagging shelves and an impossibly crowded floor. Such conditions can be expected if the employment of a full-time stores attendant is dispensed with "in view of the small size of the company." The slow movement of items is also often given as an excuse. But whatever the explanations, in most cases the real reason is a misapplied sense of economy.

PROCEDURES AND FACILITIES

The problems relating to the operation of maintenance stores are listed below. They relate in varying degrees to raw materials, supplies, tools and spare parts.

(*a*) *Organisational problems*
Staffing.
Authority to receive, issue, inspect, reject.
Responsibility for stock.
Reporting to what function?

(*b*) *Procedural problems*
Issue and receiving procedures.
Stock control methods; reordering points.
Paperwork sequence.
Dealing with scrapped, salvaged and reconditioned parts.
Identification codes of stock items.
Inventory taking.

(*c*) *Physical facilities*
Size of stores, layout, gates, mezzanines, etc.
Shelving and markings.
Lifting and transporting devices.
Safety measures, security.

Lighting, humidity control, pest control, etc.

(*d*) *Items stocked*

Supply of tools and gauges.
Grouping of materials.
Storing and preservation of spare parts.
Storing of standby units.
Standardisation.
Elimination of obsolete items.

(*e*) *Economics*

Reordering quantities.
Turnover rate of stocks.
Depreciation and interest rates.
Obsolescence.
Costing of items.
Contract suppliers.
"Make or buy" decisions.

An exhaustive discussion of all these points would take volumes and here only the more practical points can be dealt with. Several of these topics, particularly stock levels and reordering quantities can be subjected to O.R. studies. This is beyond the scope of the present work. However, the reader can profitably use this list to generate his own thoughts on the subject. Undoubtedly, all these issues need definite answers and should not be left undetermined.

To illustrate how one practice alone can affect the relationships between maintenance shops, the stores and other functions of the enterprise, let us look at the salvaging and reconditioning of parts. Whether a preventive maintenance plan is in operation or not, some parts, sub-assemblies or whole units may need to be replaced and installed. What happens to the used parts, a pump, an impeller, a housing, starter contact-shoes, a plunger piston, etc? There is a good chance that they do not go to the rubbish-heap but land on a bench or a shelf in the maintenance shop. When this practice is continued for a few months the shop is soon cluttered with discarded parts. At the back of people's minds is the thought, "Let's keep them in case we run out of stocks and we urgently need another unit." Or it may be planned to recondition these parts when "we have some time on our hands." And, of course, that time never comes. Soon work is done on the floor because benches are full and new spares get lost because they are mixed up with all the rubbish.

Obviously this is not the right solution. Those dreaded contingencies that we guard against never come and work in the shop becomes inefficient. In the course of time parts deteriorate even further, valuable space is taken up and we may even delude ourselves into thinking that we need not order new parts since we can at any time recondition some of the used ones. Whatever scrap value the parts had earlier is soon lost completely.

The procedure to be adopted is that every component should be examined as soon as it is taken off a machine. An early decision should then be taken to discard it or alternately to retain it for reconditioning. If it is to be reconditioned a work order should be issued for it thus guaranteeing that the job goes into the workload. This calls for the appropriate authorisation. Items to be disposed of, which usually constitute a majority, should be promptly evacuated from the shop area. For this we often need transportation.

When the part is finally reconditioned, "as-good-as-new," it can be stored to await re-use. And now the problems really pile up:

Is the reconditioned item to be included in the stock?
What cost centre is to be charged for the work done on it?
Where should it be stored? With brand-new stock?
To whom shall it be issued?
Is it to be issued first or kept in reserve?
When it is eventually issued, how will it be charged for?
What if a mechanic refuses to accept it?
What is the procedure in case it is issued and returned as unsuitable?
Should it appear on the stock when the inventory is taken?

These and other questions may be accounted for by the fact that few companies bother to formalise a procedure. Thus the fate of used parts is to clutter the shops. The payoff comes when in an emergency the whole place is turned upside down to find the long-discarded part, to try to recondition it in a hurry and then to find that it is beyond repair. Whatever eventual benefit is expected from hoarding parts in such a way can much more efficiently be obtained by selling the accumulated heap periodically as scrap material or to write them off as soon as they are dismantled.

"Cannibalising" of good parts from discarded machines is a simpler procedure if the residual frame or housing is to be scrapped. Otherwise it may seriously upset the accounting function in that the estimated salvage value has not been realised, whereas in fact the good components that have been taken off separately may, in fact, have been more valuable to maintenance and to the factory than the scrap value of the entire unit could ever have been.

When the central shops also manufacture spare parts, particularly in batches, other accounting problems arise. Assuming that only one pump shaft is needed for a specific repair and that the work order authorises the manufacture of five units, which jobs are to be charged with the excess units? How do we provide for a division of the original charge account number on the work order so as to make them chargeable to stores? How is the inventory to be accounted for when raw materials are depleted but home-made spares accrue?

The problem of accounting for the purchase of spares and accessories is not simple either. A frequent error is to charge the purchased stock of spares *as they are bought* on to the cost centre that they are intended for *as if they had already been consumed* and to charge them again when they are used during repairs. No wonder consumption of spares sometimes appears to be inflated. Then, of course, there is the problem of charging for a batch of bearings that have been ordered for a certain machine but may also be installed on other units.

The answer lies in charging all purchased materials, parts and supplies to the stores and to carry an inventory account. When supplies are issued the stores are credited with their value and the consuming cost centre is charged for them. The same procedure will apply to home-made parts and salvaged items.

STOCK CONTROL

The most frequently dealt with aspect of maintenance stores is that of spare parts control. A number of rule-of-thumb methods exist for reordering.

(*a*) Classify items into fast-, medium- and slow-moving groups and order new supplies at monthly, quarterly and semi-annual intervals when available residual stocks cover the same periods respectively.

(*b*) Fixed-interval, fixed-quantity orders can be placed for items that have a constant consumption rate.

(*c*) Maximum/minimum methods for more costly items requiring yearly reviews for obsolescence and turnover. Determining the two limits is not always easy and many require some statistical calculations.

When an analytical method is employed for determining the level of stocks the following factors must be taken into account:

Number of machines using the parts.
Average rate of usage.
Restocking lead time (delivery period).
Cost of tied-up capital.
Available storage space.
Cost of order (economic batch size).
Risk of obsolescence.

A common device for visualising the restocking procedure is the sawtooth diagram that represents the consumption along a time-scale and the replacement of stocks (Fig. 62). A great deal of effort has been invested in finding a fail-safe formula for determining minimum stocks. Most methods are empirical formulae relying on a stability of variables that no amount of statistical investigation can ensure. Much of it remains as conjecture, particularly since we live in an age when postal or dockers' strikes and hijacking of planes make deliveries less reliable than in the days of the horse and cart.

The effect of this state of affairs on our ability to make correct decisions is very sad, in fact we are left with guesswork. A rather effective and simple procedure that may still hold true for a majority of cases where acts of social unrest will not interfere, was published under the title: "How much to reorder when," *Factory*, McGraw-Hill, N.Y., August 1955. It simply systematises guesswork by using past experience.

The economic balance between ordering larger quantities and storing them can only partially be represented diagrammatically (Fig. 63). In this field we again have imponderables to deal with. Who can say whether machines which have all the necessary spares will not become obsolete overnight? It seems that as regards the question of stock levels, modern industry is condemned to ever-increasing inventories having reduced rates of utilisation. This is partly due to the diversification of plant and tools, to the improved maintainability of equipment and to the increasing speed of technical progress.

There is the risk of the item not being available when required. This risk and the losses involved in being out of stock disappear as ordered batch quantity is increased.

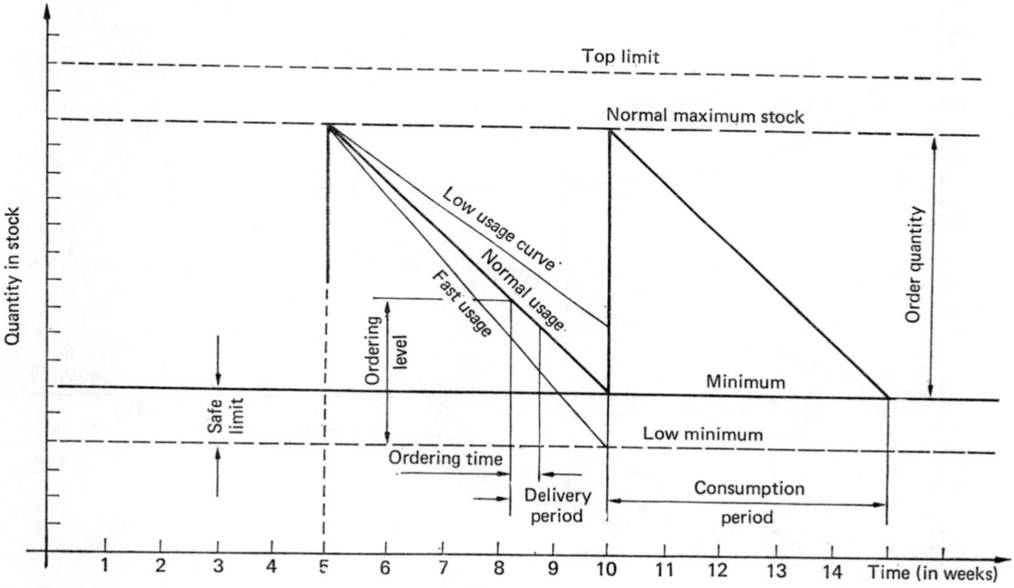

Fig. 62. Saw-tooth diagram showing depletion and replenishment of stock.

That, of course, is only part of the solution and higher stock level will increase the insurance premium. Other factors such as lack of standardisation, model changes and different catalogue numbers, affect the ready availability of spares and materials. The economies of reordering are represented in the curves in Fig. 63. To overcome obstacles of going through the "official channels" to obtain approvals to purchase or to issue, other practices will flourish.

There is, for instance, the hoarding habit of a typical maintenance man. He will hoard tools, special-sized nuts and bolts, ends of rods, bits and pieces, gaskets, washers and the lot. The trouble is that while *in his tool-box* these items are neither consumed nor used by others. The hoarding for the sake of "insurance" should be discouraged.

Another problem concerns the issue of supplies for emergency jobs. When the rush is on a craftsman will request some items before a work order has been written and sometimes even before a job number is given. For the sake of expediency the issue of essential parts should not be delayed. Thus the store attendant may issue the required items, request the recipient to sign and find out and insert the job number the next day so as not to hold up the completion of urgent jobs. Issue slips without a job number may cause worry to the accountants and such friction should be avoided whenever possible—after all we have to coexist.

Considerable cost savings can be achieved when workers take the trouble to return remnants to stores. This practice will, however, increase costing and accounting procedures. To be able to issue only such quantities as are required for the job, the stores

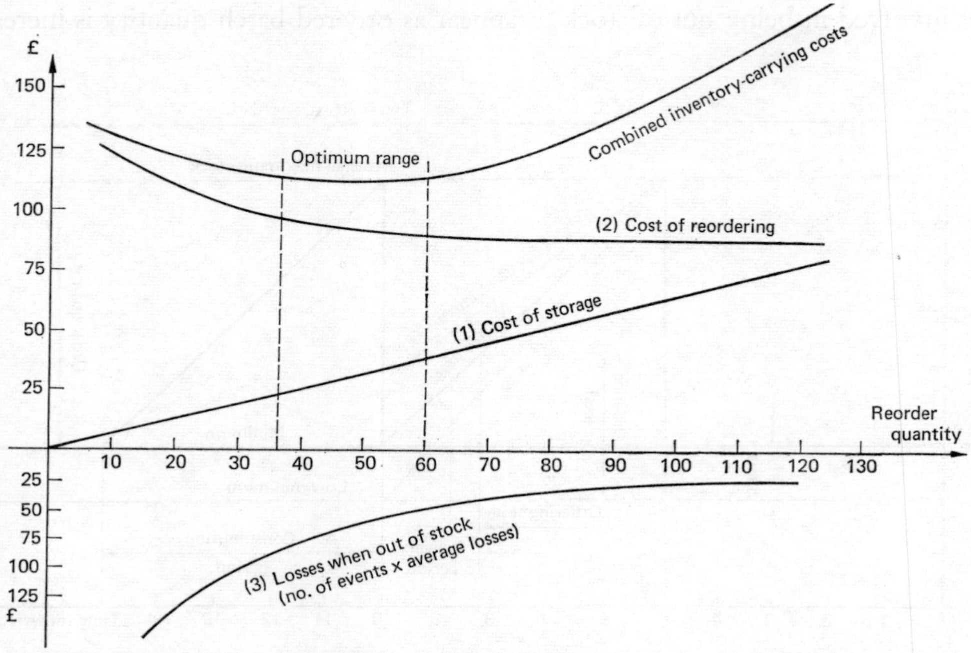

Fig. 63. The effect of relevant factors when considering the economic balance of stock levels.

should have a selection of cutting tools. Supervisors can contribute to the effective use of materials by planning jobs in advance and requesting specified quantities of materials.

Stores procedures and practices should allow quick and efficient service. The objective is to minimise the time required for making issue slips, getting them signed, walking to stores and queuing at windows. Some of the measures that can be adopted are:

(*a*) Open bins for standard sizes of nuts, bolts and washers, to be placed next to stores windows. This will probably reduce the incidence of queuing by approximately 30 per cent.

(*b*) Sub-stores on mobile racks carrying a selection of frequently used expendable items to be placed at points of use and replenished periodically. This will reduce walking time.

(*c*) Marking and identification of shelves inside the stores and a systematic division of items so as to allow them to be easily located.

(*d*) Pre-tagging of costly items showing identification, cost of unit and plant unit for which they are intended. Upon issuing such an item the tag is retained and attached to the issue slip, so as to minimise writing time.

(*e*) Pre-printing of issue slips for standard items so that only date, quantity and job number have to be filled in later. The slips are kept in a pocket attached to the shelf. This can be standard practice when a computerised stock control is adopted.

(*f*) Pre-selecting a set of tools and parts on trays for certain recurring jobs, or pre-packaging sets in nylon bags.

(*g*) Revolving "lazy-susans" for small items saves walking time of store attendant when placed next to issue window.

(*h*) Displays of standard items and their code numbers on a wall-board, such as wood-screws, cables, etc. so that they can be requested by code numbers.

(*i*) "Self-service," supermarket style, may be introduced for all items whose issue has to be recorded, with the exception of expensive tools and parts. These can be kept locked and issued by attendants. Items are then checked and recorded at the exit gate.

ASPECTS OF CONTROL AND COMPUTERISATION

Stores control relates mainly to (*i*) keeping of adequate stocks and replenishing them and (*ii*) accounting for the issue of items to jobs. We have so far discussed measures leading to effective service. Certain procedures were suggested that both satisfy the requirements of expediency in issuing and at the same time do not neglect the recording aspects. What are the necessary procedures to realise the objectives of a maintenance stores control?

Let us assume that the basic elements are available and operated *manually*:

Bin cards, to record issues as they take place, and to facilitate the taking of inventory.

Stock cards, to record balance in stock and balance of orders placed, and maximum/minimum levels.

Stores request slips, to list required items, quantities, job number or charge account, and person receiving items.

Purchase requests, to indicate date and quantities ordered to replenish stocks.

Charge accounts, or cost centres "consuming" maintenance work, to which stores issues are charged. Their coding system should be as simple as possible and yet complete.

Delivery notes, arriving at stores with the goods, to allow accounting with suppliers.

Invoices, by suppliers, indicating cost per unit to which handling expenses may have to be added.

When the stores control operation is on a small scale a manual method involving a certain degree of inconvenience is tolerable. However, on a larger scale the inefficiencies become too costly.

At this stage management might well start looking for ways to streamline this activity. While it is true that there are intermediary stages leading to computerisation, half-measures are to be avoided. Often tabulating machines are purchased and these may satisfy the needs for five to eight years. But sometimes due to speedy increases in volume of work they may prove inadequate and then again faster equipment may be necessary. Mere improvisation in such cases is bound to prove unsatisfactory.

With a manual system adequate stocks can only be ensured by visual checks of shelves, bin cards and/or stock cards. When this erratic check fails, maintenance supervisors will occasionally inspect the shelves themselves. Inventory taking is laborious and so is compiling monthly expenditure for all charge accounts. Request slips take a long time to write, they may get lost and other mishaps occur. Accounting for a large number of jobs, charge accounts and cost centres becomes a heavy burden. When all these symptoms occur at the same time the idea of computerising seems very appealing.

There is no doubt that when an operation reaches certain proportions we can turn to electronic data processing (EDP) expecting good effects and economy. To define "certain proportion" is not easy, but the following may serve as a guide:

A maintenance team of over fifty.
An installation valued at over £800,000.
Two to three hundred charge accounts.
A monthly turnover of 300 or more jobs.
Five thousand or more items in stock.
Value of stock in excess of £100,000.

(It should be borne in mind that with the steady decrease in the price of electronic equipment and commercial data processing these figures will also gradually diminish.)

The time and effort required to introduce EDP may vary with the prevailing conditions, the choice of system and the size of the project. Changing a *good* manual system to a computerised one is only a matter of new forms and an investment in hardware. However, in practice it will be found that organisations are often not ready for EDP. Certain procedural sequences are not followed, there are no numbering codes for stock items, charge accounts, plant and equipment, workers, etc. Since EDP relies on coded information these are essential prerequisites. If we wish, for instance, to get a print-out of short items below minimum level, this level must be predetermined on all items and coded, which is often a very laborious chore.

Considering the use of electronic or mechanical methods for stores control and job accounting there are the following basic alternatives:

1. The use of tabulating or accounting machines (these are electromechanical devices).
2. The introduction of punched cards or tape which will be processed by computer bureaux.
3. The use of visible-record computers (these are electronic calculators and accounting machines).
4. A combination of these methods.

Tabulating machines have been brought to a high level of sophistication by the major companies (NCR, Burroughs, Sumlock, Singer and others). They can post entries, calculate balances, multiply quantities by unit cost, print out subtotals, totals and grand totals. Some can be electronically connected to other mechanical devices for collating, posting and adding data. The best purpose to which tabulators can be put is therefore the posting *of stores movements on stock cards* and the charging of job accounts.

The punched-card computer on the other hand is used for marshalling data into different classifications, while performing any number of calculations. The results are then tabled in a print-out in any preferred sequence. While accounting machines cannot do this (except in a very limited way, for very limited periods), it must be accepted that *the computer cannot post data.* It cannot post an entry as it occurs on individual stock cards, although it will give a print-out at the end of the period, listing the balances. Barring very expensive installations, which would include disc drives and keyboard terminals, a computer cannot handle single operations and it must wait for the end of certain periods for a group of operations to be processed. In between two reports one is without information, and again when the report arrives (at last) it is already out of date.

Although these remarks may provoke strong criticism from advocates of the computer and their salesmen, certain operations are feasible but are simply not economical for the purpose discussed here.

Since tabulating machines and computers obviously *complement* each other, a combined system is here suggested. In simple words:

(*a*) While the tabulating machine posts the entries on stock cards, job cards and accounts, and while these postings are keeping the cards up to date, the tabulating machine is connected to a punching device, where

(*b*) a card or tape is punched for each entry. These are periodically sent for print-out, a process which regroups the data by departments, or by any other requested sequence.

Thus at any time stock cards are kept up to date by the accounting machine on all issues, orders and balances. The computer serves as a classifying and summarising tool for monthly returns.

As a by-product the accounting machine automatically indicates items to be ordered. This is done by the operator during her normal procedures for making the entries. All she has to do is read the minimum level from the card that she is handling, and type it out on the keys. If the balance after issue is less than minimum, the machine indicates it by a red entry at the end of the line. At the end of the day, the sheets containing all the transactions for the day carry a list of items to be recorded. The programming bar (or its equivalent) can handle such and similar procedures within a very wide range.

Competitively speaking, a computer can also do this by means of the cards (or tape) that were punched at the same time and it can be programmed to initiate purchase orders but we have to wait until the end of the month when it gets batches of forms for processing. When the balance of an item comes below the level as a result of an issue, the computer uses the master card to read off the minimum level and to show the missing balance and print out a purchase request slip for a predetermined quantity, supplier, cost per unit, total for order, expected date of delivery and more if necessary.

The operation of an accounting machine linked to a punched-tape device, plus the cost of outside data processing would not exceed £400 per month including the salary of one operator and all the reports you can read.

There are cases when the company is already operating a computer for other procedures, such as payroll, sales or production control. If so, the added procedure is not too difficult to introduce. However, sometimes administration of the production and payroll does not warrant computerisation, while maintenance stores control does justify it. A typical case in point is a mine or chemical plant, where the value of plant imposes a heavy maintenance expenditure, while payroll is relatively easy to manage.

When the company does not own a computer, which is sensible for small to medium companies, the data can be punched and checked on the premises on rented equipment, and sent out for processing to a data centre. Alternatively, you can link your machine, a terminal, to a DP bureau on a time-sharing basis and thereby be up to date at all times. This would be termed as being "on-line," and you could operate on batch-mode (update accumulated data at intervals as required), or real-time. The latter system would update instantaneously as events occur.

Computerisation, to be sure, imposes a degree of "system dictatorship" which is not easy to accept at the start. It takes staff some time to realise that the earlier informality of "Hello Bill, we need some bearings in a hurry, old boy," which incidentally also gave rise to quite a bit of friction is now a thing of the past. Procedures and sequences must be adhered to with a computer and above all coding must be respected.

What often bothers unwilling converts to EDP is the fact that they cannot see the data in tangible form or that they cannot visually follow the computations. This is called the "black-box" effect. Some justified criticism is also levelled against late delivery of reports but as long as we insist that accounts are closed on the last day of the month, delays in the appearance of some of the reports is unavoidable. Tabulating machines and VRCs, of course, side-step this issue. They are up to date as you go.

If a maintenance manager were allowed to choose between a tabulating machine and a computer, he would be wise to choose the former. Since the emphasis is on immediate and up-to-date information, a tabulating machine can provide it more economically. One machine can handle up to 160 entries per hour and if volume increases, two shifts of seven hours can be worked. In most cases, 1120 entries a day should be adequate.

A typical problem that arises when EDP is introduced is the numbering system of items. The first hurdle is that of grouping items into materials, parts and supplies. Then the next hurdle appears. The numbering catalogue must be kept up to date with deletions, new entries and changes. There are many facets to this problem which is at its worst in the field of spare parts.

Since manufacturers have their own numbering systems, the part will now bear two multi-digit numbers for complete identification, the company's and the manufacturer's.

Overcoming this problem requires patience. One way to do it, is to establish a number code for each manufacturer and prefix it to their number. There is one snag in this: spares in this group must be preceded by one digit that will set them apart from all the rest of the stores, *e.g.* the digit 9 which would indicate that they are manufacturers' spares. A two-digit number will identify the manufacturer. Other distinctions may be necessary such as mechanical spares as against electrical parts.

9.12 xxxxx—Square D
9.27 xxxxx—Rockwell
9.31 xxxxx—Cincinnati
9.58 xxxxx—Ingersoll–Rand, etc.

Other problems include that of spare parts that are interchangeable with different manufacturers' machines or identical parts that the manufacturers renumber for their *new* models. Combinations of these and similar problems can go on *ad infinitum* to the chagrin of programmers and systems staff.

Still another puzzle is the treatment of occasional parts that are not normally carried in stock. These are usually issued as soon as they arrive. Perhaps the best way is to prefix these items with the digits 99 and to assign a number made up of the date and the serial number of the "goods-in notice." Since copies of these are kept and later used in calculating the price when the invoice is received, the reference to this number would identify the item at will.

To be sure, the introduction of EDP is not without difficulties. Among the greatest obstacles is the human factor, namely getting people to conform to requirements and getting their wholehearted co-operation. Nevertheless many benefits can be derived from the use of computerised EDP as suggested below. Data that appear on work orders and stores issue slips allow the following arrangements of print-out:

(*a*) List of jobs worked on during a period, man-hours and materials issued for each of them and total *cost of each job.*
(*b*) *Work done in each department* and its total costs.
(*c*) List of *plant units and the monthly cost* of servicing for each.
(*d*) List of all issues per *group of items* and their cost.

There is a tendency in the early days of computerisation to insist on too many such reports. Some of these will later prove to be unnecessary. Thorough planning will help in avoiding such pitfalls.

The real advantage that EDP offers is in the handling of a great amount of data in a very fast and constant way. Once the system is operating, all normal work processes are reduced to the vetting and input of data and only exceptional problems require personal attention.

Other advantages that accrue from EDP are in inventory taking and yearly review of stock turnover. Items that do not show movement year after year can easily be identified and eliminated. In fact, if planned in that way, the computer can provide a complete print-out of the balances on all stock items at any time during the year.

AN APPROACH TO REPLACEMENT PROCEDURES

This aspect of plant engineering is nowadays closely scrutinised by principles of "maintainability" which have sprouted from the efforts of the U.S. Department of Defense since 1954 to rationalise their purchase of military equipment. These principles endeavour to optimise the cost of items as against their throw-away value.[2] Naturally, many ideas are only applicable on a scale of operation approaching that of the D.o.D. We can therefore do little in this respect except take note of the concept and raise the issue with equipment manufacturers provided we can put enough weight behind our arguments.

On a more down-to-earth level, replacement problems of plant and equipment relate to the following three managerial functions:

(*a*) PRODUCTION—The users of equipment.
(*b*) MANAGEMENT—The providers of money.
(*c*) MAINTENANCE—The advisers on replacement.

Our discussion here relates to the role that the maintenance function should play in the decision-making process when replacements are contemplated. A good maintenance system will enable maintenance managers to fulfil the role exceptionally well since providing the relevant data is an important by-product of the service. The question of replacements to satisfy the expansion of business volume is here deliberately avoided. In such instances maintenance costs are not the deciding factors involved in the decision. Let us, therefore, discuss here the instances relating to replacements as a direct outcome of mechanical deterioration in various aspects: wear and tear, inaccuracy and lack of uniformity of products, low degree of reliability, high frequency of stoppages, increased safety hazards, excessive consumption of power and rising maintenance costs.

As the age of a machine advances other problems are encountered. Difficulties in finding the required spares, lessening chances for obtaining a purchaser, integration of old machines with newer ones in the same production process and finally a poor appearance. Pressures exist to reduce the amount of bother that old machines cause and to counteract the competitive strength of new machines in the hands of a competitor.

Technical obsolescence affects output capacity and performance in a comparative manner that changes as time passes. Such obsolescence is greater the faster the rate at which technological changes are introduced and absorbed in industry. Newer machines are usually much more complex and more costly to maintain in spite of such improvements as central lubrication systems, long-lasting bearings and the use of new and better materials. Thus "replacement" is not strictly the right term to be used. Only in few instances will a new machine be similar to the one that it replaces.

Various approaches have been made with a view to present a usable formula for replacements. In recent years operational research techniques have been applied to replacement problems. Although that approach may produce numerical values as solutions, the underlying considerations are often left unresolved. In the end it appears that there is still no substitute for experience and some common-sense. A review will show that the proposals hinge on the balance of groups of factors that are considered. The comparison of profitability can be approached in several ways:

(*a*) The proposed new machine's ability to produce more profits than the old one.
(*b*) The rate of return on newly invested capital.
(*c*) The cost of operating the new machines as compared with the old ones.

It is in the last procedure that the maintenance function can contribute most, by answering the following questions:

(*a*) *The existing machine*
  (*i*) How much trouble is it giving? What has been the frequency of stoppages and parts replacements required over the past few years?
  (*ii*) Is there a definite trend or pattern of deterioration evident? Has the consumption of parts and the cost of repairs risen significantly in recent months?
  (*iii*) What amount of effort has been spent on maintaining the old machine?
  (*iv*) What are the chances of getting the old machine into good shape again and for how long would that be good?
  (*v*) Is the old machine good enough to be sold at a reasonable price?
(*b*) *The proposed machine*
  (*i*) Do we have the know-how for its upkeep?
  (*ii*) Will it require any specialised services, parts or skills?
  (*iii*) What is the assessment of its expected trouble-free performance?

Once management is in possession of this information in a quantitative form a decision can be made incorporating the aspects of *available finances, competitive position* and *operating efficiency*. The by now famous MAPI Formula,*[14] employing a useful mix of maths and clear-cut policies, provides a useful approach as it endeavours to take into account the first and last aspects, since nothing much can be done to stop competitors from competing. However, we can improve our position by using machines and capital effectively.

According to the formula, the capital cost on an investment decreases as time passes, however, the "operating inferiority" of a machine increases. This means that with the passage of time we can cheaply use an outdated machine. The combined cost of capital and operating expenditure produces a curve which presents a low point somewhere along a time-scale. The calculation centres round this "adverse minimum" which determines the inferiority of the old machine, "the defender," as opposed to the projected adverse minimum of a proposed machine, "the challenger." A great many assumptions and estimates have to be made.

* The original version has since been revised and is now fully discussed in a book entitled *Business Investment Management* published by the Machinery and Allied Products Institute and the Council for Technological Advancement, 1200 18th Street, Washington, D.C. 20036, U.S.A.

In spite of the range of factors that the formula takes into account, it can neither "encompass the infinite variety and complexity encountered in practice, nor can it be a substitute for sound judgment," as expressed by the authors themselves.

Procedures compare "defenders" and "challengers" by a set of factors that can be detailed over several pages of print.

(*a*) Direct (operating) labour costs.
(*b*) Supervisory and administrative costs.
(*c*) Indirect (services) labour costs.
(*d*) Maintenance costs.
(*e*) Cost of supplies (consumable).
(*f*) Cost of power.
(*g*) Cost of space, taxes and insurance.
(*h*) Quality figures and reliability.
(*i*) Output of performance.
(*j*) Design details for operating convenience and appearance.
(*k*) Ease of servicing.
(*l*) Proximity of agents.

The last three are difficult to quantify and are sometimes added in the form of remarks at the bottom of the form.

A number of alternative methods have been proposed and practised by leading industrial companies. A selection of forms utilised by them appears in Appendix III of this book. They illustrate the practical down-to-each approach that is often preferred to complex financial theories. The validity of these procedures can only be proven by the fact that all these companies so far have stayed in business.

It again becomes clear that maintenance can play an advisory role in only a few of these aspects. It is suggested that a systematic appraisal of machines is preferable to the use of impressions and hunches. The maintenance manager or supervisor should attempt to formalise an approach which will help him reach conclusions in a more consistent way than is usually done today. An annual survey of equipment offers a good opportunity to sharpen one's ability to assess replacement problems.

Chapter 8

Analytical techniques for improvement

All too often companies become over-dependent on the personal performances of two or three members of the maintenance team. It is apparent to onlookers that it is only the experience and devotion of these few that keep the show going. In times of crisis they get things done and when information is required they seem to know it all; they have it all at their fingertips.

But this situation creates its own problems. When for some reason or other they are away from work everything goes with them. Indispensable though they appear to be, the company must find ways of carrying on without them. The very fact that they know their job so well must mean that most innovations have passed them by. To be sure, *they* know the job inside out as nobody else does; but accepting their working methods may quite easily lead to a sort of mental inertia where nobody questions *why* things have to be done in a certain way and why costs and results are as we find them.

Since time inexorably marches on, new machines are installed, new methods are introduced and the company requires more and better data to manage its business. It is for this reason that management cannot leave matters as they are. Thus, the existing ways must be investigated impersonally from several angles in order to discover possible improvements and to discover the factors that may lead to greater efficiency.

Analytical techniques provide the means whereby we get a closer insight into the ways in which a service is implemented. By applying a variety of analytical techniques we ensure that results are corroborated from all angles. These will indicate how to deal with problems. Simple procedures can sometimes lead us to recognise significant facts or may lead us to realise that seemingly disoriented facts fall into a relevant pattern. Thus we can interpret events, facts and figures, and provide bases for improving results.

A great many analytical techniques that lie within the field of industrial management have been developed in recent years and it is probably only a lack of time and resources that prevent us from applying all of them profitably to maintenance operations. The application of O.R. techniques to maintenance is well covered in a recent book entitled *Operational Research in Maintenance*,[6] which deals with replacement strategies, overhaul policies, size of service crews, stockholding decisions and multi-project planning.

In keeping with O.R. theory all these topics are orientated towards finding optimum economic solutions by means of mathematical equations. However, for O.R. techniques to provide tangible results the scale of operations has to be above a certain size in order to provide adequate data. The same condition will also allow it to become financially viable particularly since the costs of such studies may be considerable.

The material presented in the present volume is meant to be implemented by more modest efforts that can often be mustered within the maintenance department itself. The effects may also be of a more immediate nature.

The analytical techniques presented in this chapter concentrate on *methods* of working, on the *time* it takes to do the work and on the investigation of *occurrences* that upset normal work. Since this book has also taken a close look at the *procedures* employed in performing the maintenance service and contains chapters on the *economics and controls* of maintenance, we can claim to have covered all important aspects. By applying these techniques we can ensure a more thorough implementation of the "loop of action" concept which was presented in Chapter 4.

WORK STUDY IN MAINTENANCE

Many aspects of maintenance lend themselves to work study and improvement. It is not enough to pay lip-service to the phrase "There's always room for improvement," active steps have to be taken to get closer to perfection. The most helpful part of work study is its potential for analysis. By getting the facts of a situation and by quantifying its variables, work study becomes a problem-solving tool and invaluable in trouble-shooting.

It would be wrong to think of work study as being mainly work measurement, leading to job times and incentives. Time-based incentives are still not as widespread as its advocates would lead us to believe; neither are they the panacea they are proclaimed to be. Time study and work sampling should be used in maintenance, but their objectives should not be limited to time standards (incentives are discussed in Chapter 9, pp. 176–201).

It is equally wrong to decide upon a cost-reduction campaign and embark on a spurt of work study to "reduce costs by 15 per cent over the next three months, please." Work study should be applied at a steady pace, planning for consecutive surveys covering all activities, presenting results and making recommendations all along. It is remarkable how often suggestions by maintenance staff are ignored, only to be "discovered" when a study is made. The point here is that, for example, tradesmen's tools should be updated and made more plentiful. This and other similar suggestions had, of course, been proposed earlier and ignored. In fact the search for efficiency and the ability to overcome problems is inherent in workers who grapple with different jobs daily, although they may not pursue them methodically. Work study provides the systematic–analytic approach that helps to concentrate efforts on the more important issues. At the same time the work-study approach serves as a catalyst between line workers and management.

All the "tricks of the trade" that work study has to offer are welcome in maintenance, including layout studies, motion study, work-bench layout, materials handling, work sampling, etc. We can use these techniques in the tool room, spares stores, lubrication service, mechanical and electrical workshops and on location wherever tradesmen carry out their work. But where do we start, and how?

Since time is a prime commodity, the efficiency of the department revolves round the *effective time utilisation of its workforce*. We may ask, for instance, what happens from the moment a job is assigned to a mechanic until the job is completed? How much of a craftsman's time is effectively used? How promptly can a craftsman obtain tools from the stores? How much time is wasted locating people and getting requisites when a breakdown is reported? Or when a steam-pipe leak suddenly spills the gasket how long does it take to replace it? How are jobs assigned when workers report for work in the mornings? Why was the last job interrupted three times? . . . and so forth.

It would be helpful to have replies to these and other questions. It would be most satisfactory to take the sting out of such provocative queries. This is the field where work study can make a contribution. For example, from the moment an electrical failure is reported in a department to the time the machines are again in operation many things may happen. The less efficient the sequence of events, the more there is to be improved. The sequence of any work can be divided into three stages, namely:

PREPARE—DO—PUT AWAY.

Within these stages a wide variety of unproductive elements occur: waiting, walking, searching, setting up, preparing materials, parts and tools, etc. and finally cleaning up and returning to base. The actual repair time may have been as little as 20 per cent of total time elapsed since the failure was reported. These are some of the findings of investigations carried out over a span of two decades.

A lot of wasted time results from lack of contact between the different craftsmen. Foremen have to be found and consulted, the despatcher has to issue the work order, the tool-room attendant and the stores issue clerks have to find whatever is required and the supervisor of the requesting shop as well as the nearest operators have to be consulted.

The search element is an ever-recurring one. It relates to people, tools, materials, location, causes of failure, getting assistance, transportation, etc.

Identifying causes of failure is a very time-consuming process, and while work study may be of some help it is much more effective to train people in systematic fault-finding techniques.

Other untoward events may hold up work. The tradesman's tools may be in poor condition or they may be inadequate for the job. Or the job may need a special tool that has to be either purchased in a hurry or improvised on the spot.

Equally important is the availability of spare parts as prescribed in the manuals and, even when they are available, locating them in the stores also takes time.

Sometimes faults that require attention on installations of pipe-work or large plant units are difficult to reach, and some corners may have to be cleared or scaffolding may have to be erected before any work can be done.

Finally, the mechanic or electrician must be able to handle his job properly. His ability may range from earlier experience with the same machine to total lack of previous contact with it. What happens in the latter case depends on his skill in interpreting the manual, if there is one, or on his individual talent and ingenuity. Training thus plays an important role in achieving better time utilisation.

The list of delaying factors is inexhaustible and we never know where and how they will appear. There is hardly ever a smooth sequence. However, approaching the ideal of uninterrupted work is the challenge presented to work-study analysts. If we don't want the unforeseen to catch us unprepared, we must at least try and minimise its impact.

When jobs are to a certain degree repetitive, such as lubrication and inspection, there will be no work-study problem. Overhauls too can be carefully sequenced and jobs requiring the co-operation of teams can also be scheduled. The way to handle emergency calls can be rehearsed and pre-specified. Thus, a list of areas in which work study can be useful would include some of the following:

(*a*) Standards for *repetitive jobs*, namely, normal times and work specifications.
(*b*) Tradesmen's tooling and the *range of tools* available in central tool stores.
(*c*) *Standardisation* of equipment, spares, lubricants and materials.
(*d*) *Handling methods* and means of transportation.
(*e*) *Teamwork planning* based on timing of work.
(*f*) *Pre-locating of spares, e.g.* fuses, gaskets, belts, etc., at point-of-use.
(*g*) *Precautionary measures*, such as safety devices, colour code and identification markings.
(*h*) *Layout* of workshops and work-benches.
(*i*) Methods of *communication*.
(*j*) *Paperwork* procedures.

The above points can also fruitfully serve as targets in cost-reduction efforts. They all represent cost components and their potential for achieving savings is great.

Let us examine some work-study techniques that may serve as fact finders and which at the same time may help in the solving of problems.

Production studies

Assuming that in maintenance there are few jobs which re-occur in the same form, the application of time study over a wide field is not feasible. Instead, a specific branch of time studies, namely production studies, is used. In work study this term refers to a *recording of details and a continuous timing of an observed sequence* with a view to identifying effective work elements. The stripping down for cleaning and reassembly of a spinning-frame would be a good case for the application of this technique. Another example is the performance of a routine inspection of a railway locomotive.

When, on the other hand, a production study is made on the departmental supervisor's activities and that of his foremen, we really get to grips with managerial problems. One such study showed that craft supervisors spent only 25 per cent or less of their time on giving instructions or inspecting work but over 20 per cent in walking and searching. Enquiring about people, spares, tools and the equipment to be serviced, amounted to about another 25 per cent. Clerical work of all sorts including reports and correspondence constituted no more than 10 per cent and the rest of the time (20 per cent) was spent on co-ordinating with other people. It is to be queried whether walking and enquiries should take up about half of a supervisor's time.

A frequency analysis of the contacts with people usually discloses a low degree of planning and poor communication. When dealing with subordinates, matters relating to "Who is on afternoon shift today? Do you know where Mike is today? Can you do overtime tonight, Jack?" are discussed. Contacts with production shop supervisors are mostly concerned with replying to complaints or in trying to convince a production supervisor of the necessity to do the servicing of a machine next spring. Contacts with management too often consist of pleading for spares, for extra manpower or for approval of a budget to improve the workshops.

The interesting fact about these results is both in the nature and frequency of these contacts, in the topics discussed and also in the time spent on them at the expense of more constructive work.

Production studies also disclose causes of delays and sources of irritation. Such studies can be oriented to particular aspects of maintenance work. Individual tool-chests and the department's complement of tools can also profitably be surveyed from time to time. A check-up on the way plant history information is recorded and analysed is another example. Planning of work during an approaching shutdown is often postponed until it becomes a crash programme. When such diverse activities are investigated they indicate the reasons for poor performance.

Once the reasons for delays are identified, the remedy lies in the application of measures that soon become routine and they speak for themselves. For example a peg board for planning shift work can eliminate much talking and questioning. A reliable memo system for notifications of all sorts, *e.g.* "the parts of XYZ have arrived," or "your B-line is due for service next Wednesday," etc. can appreciably reduce the need for enquiries. Emergency calls should possibly be handled by light signals, buzzers or other devices, and communications between people can be modernised by installing any one of the large number of modern electronic devices. Locating workers and communicating with them represents perhaps the greatest waste of time for supervisors in most maintenance departments.

Production studies go a long way in assessing the extent of problems and their importance, and may well indicate measures that can be taken for improving the situation. Although it is a relatively costly device it is not prohibitive, in the same way that direct time study is not too costly for production work. In recent years, however, work sampling (*see* p. 168) has been increasingly applied in preference to production studies. Nevertheless, the latter still remains an effective tool in the study of prolonged job sequences.

For the purpose of establishing a lubrication routine there is hardly a better tool than production studies. When a lubricator is studied in his daily rounds, usually a not very well-organised job, a production study would show the effective and ineffective time elements and their percentages. A grouping of elemental times will enable average values to be calculated for effective work, for unavoidable incidental elements such as refilling of greasers and oil-containers and for walking times. Accurate frequencies of recurring elements can also be established from the studies and thus several days of production studies can provide enough elements to allow the synthetic compilation of times for this task. Naturally, we would need all the relevant data, such as shop layout, a list of machines, points to be lubricated, types of lubricants, etc.

Production studies can also be applied to individual jobs which are costly in man-power and re-occur in seasonal frequencies. From such studies an optimum sequence for the job can be developed and in some cases Gantt charts can be drawn up to further increase work effectiveness.

Layout and flow diagram

These techniques would apply mainly to the examination of layouts in repair shops or movements of the storekeeper inside the tool and spares stores.

The existence of layout blueprints for all areas where plant and equipment are installed should be considered vital. It facilitates planning of work, issuing of instructions, establishment of lubrication and inspection rounds and the division of duties for decentralised staff.

Another use for layouts and flow diagrams is as a guide in the calculation of job times where estimates are required. A typical example would lie in compiling an estimate for a repainting job involving several buildings, a certain amount of pipe work and some steel construction. In the absence of layouts an analytical estimate would be impossible.

Layouts and flow diagrams also prove their worth in saving walking time. In conjunction with production studies they could show for example whether strategically situated sub-stores would reduce walking time. Location of the maintenance shops in relation to the main production shops would also thereby be highlighted and this could easily prove whether relocating the shops would be justified.

Another application of layout principles is in the positioning of work benches. It is a fact that in most maintenance shops very few work benches are actually used for working upon. They are too often piled high with discarded components, replacement parts and a variety of tools. It is pathetic to what extent work benches become repositories and storage shelves for junk. As a result work has often to be performed on the floor under very trying conditions. Soon the floors also become cluttered with parts of machinery. In some shops the doorway is the last spot where you can take a straight step.

The establishment of proper workplaces, well equipped with a supply of consumable components, nuts, belts, washers, etc. and a good selection of tools, is an important step towards efficiency. Many repair shops would also benefit greatly by the provision of a few large scrap-disposal boxes. Work bays which incorporate the principles of motion study and layout are essential for the maintenance of vehicles, rolling stock and the overhaul of machine tools.

A direct outcome of layout analysis is the discovery of handling problems. When the movement of machines, parts or tools or people is investigated it soon becomes clear whether the methods for moving them are efficient or not. When long distances or different levels of the building are involved, handling methods become even more important. To put it bluntly, there is no reason why skilled people should waste their time and effort in attempting to overcome poor handling methods and poor layouts.

Cost-reduction checklist

The following checklist mentions the most fertile areas for time saving and cost reduction. Although most of the points are discussed in length throughout this and other chapters, it is useful to list them here in a concise form.

(*a*) *Maintenance shops*
- A sensible layout of the shops.
- Adequate shelving and storage areas.
- Handling equipment for lifting and moving heavy components.
- Satisfactory tooling for work performed within the shops.
- Adequate individual tooling for craftsmen.
- Proper work benches and correct workplace layout.
- Adequate lighting and ventilation.

(*b*) *Manpower utilisation*
- Supervisor's desk properly equipped to perform his duties.
- Work-planning boards for visual control.
- Correct job assignment procedures.
- Decentralised assignments where justified.
- Satisfactory daily time-card procedures.
- Foreman's daily summary of hours.
- Analysis of time spent on clerical work.
- Effective communication systems (*see* following section).
- Training and upgrading of craftsmen.
- Training of helpers and apprentices.

(*c*) *Service implementation*
- A proper work-order system.
- Adequate recording procedures.
- Service schedules for all repetitive jobs.
- Standard instruction procedures.
- Periodic surveys of effectiveness.
- Regular co-ordinating sessions with production management and quality control functions.
- Standardisation of lubricants, materials and parts.

(*d*) *Control procedures*
Regular reporting procedures.
An effective supervisory staff.
Periodical planning meetings for crafts foremen.

(*e*) *Work details*
Pre-positioning of replacement parts at point-of-use.
Prefabrication of replacements.
Immediate reconditioning of re-usable parts.
Accessibility of service points.
Minimising the effects of naturally occurring detrimental conditions, *e.g.* hard water, abrasive dust, water in compressed air, corrosive agents, etc.
Prompt discarding of junk.

Communication media

Proper and adequate means of communication are of particular importance in maintenance. Tradesmen who are usually dispersed over a wide area cannot effectively perform their work unless all their requirements converge at the right time and in the right place. Not only do we need specified materials, tools and supplies, but people of various levels and responsible to different authorities have to agree to certain things taking place at a certain time. To achieve this every day of the year is quite a task. Workers must be supervised while they are performing their tasks, their work must be inspected and sometimes assistance must be given and this requires that the supervisor be able to locate his men at all times. No wonder maintenance managers and supervisors spend on the average close to 50 per cent of their time in talking, walking, searching and explaining. The communication problem not only relates to contact within the department but also to contact between maintenance supervisors and foremen of other functions.

Ways and means must be found to minimise both the time and the nervous strain entailed in co-ordinating these efforts. In the normal course of events too much is left to chance communications and this gives rise to frequent disruptions. The safest way would be to set up a continuous and reliable system for channelling information from the various sources to their different destinations.

We here distinguish between ORAL, WRITTEN or ELECTRO-MECHANICAL means of communication and the best of these should be applied in maintenance. Oral communication between two people is often unreliable and when an exchange of opinions is required between three or more people it can cause serious misunderstandings. For this reason *factual* information, *decisions* and *dates* should always be communicated in *written form.* When meetings take place for the purpose of mutual briefing, circulation of agendas and subsequent circulation of notes on the meeting is essential. However, planning and decision-taking sessions should always be followed up with circulation of minutes.

Oral communication, *i.e.* planning meetings, is necessary since in many cases obtaining agreement is essential. The purpose of such meetings should be explicitly stated in advance so that people can come prepared. Regular fortnightly or monthly meetings are advisable in most companies of medium size.

Written communications are often considered a waste of time. Although there is no safeguard against occasional inefficiencies, ways can be found to make both the writing of instructions and their interpretation more effective. Where paperwork is not too well developed, it is often supplemented by improvised notes, memos and slips of paper. It is advisable to introduce forms on which certain standard communications are pre-printed and all that is needed is to check them off where applicable or to fill in a few missing words, as for example: parts for ...... have arrived, no authorisation yet for ......, please note service of machine No. ...... due on ......, etc.

Notice-boards are another means of conveying written information. These are often ignored owing to the poor way in which they are utilised. Notice-boards should be divided into specific areas for such subjects as technical information, leave and shift-work, personal notices and social events. They can also be used in production shops to show the weekly schedule for machines to be serviced. Forms attached to the boards should be distinctive in colour and shape so as to allow easy identification.

The way messages are transmitted often give rise to delays and oversights. Within the maintenance department, all key personnel should have definite IN-trays at pre-determined spots.

Regular daily and weekly schedules of the maintenance manager and his staff of supervisors have a great influence on the reliability of communication. Key personnel should establish regular times for meetings, for seeing maintenance staff in the office, for reading communications and for making the rounds of the shops. Although this advice appears to be a Utopian wish, we all know of certain managers who can always be relied upon to do certain things at certain times and we realise how much confidence such knowledge inspires. How much time could be saved in trying to contact people by phone, if we knew that "He usually goes to an inspection between ten and eleven-thirty?" It would equally reduce lost time if workers knew that on personal problems the foreman is available daily between three and four in the afternoon, or the storeman can expect the supervisor's visit at eleven forty-five in the morning.

That written communication imposes a burden on the writer is unquestionable. However, if done effectively written material can be helpful to everybody, including the writer. Keeping a diary is useful for preparing reports. Reports are necessary for summarising activities periodically. And summaries are essential for interpreting data and making decisions. It is safe to state that people who take notes and who refer to them often find them invaluable. Since memory is limited in capacity and permanence, note-taking appears to be unavoidable (the use and purpose of reports is discussed in a subsequent chapter).

While the use of the latest electronic devices, such as closed-circuit television, are occasionally justified, there is also a place for such simple devices as buzzers and coloured blinking lights. In one instance a simple two-coloured plate solved a communication problem. This happened in the following way. A certain large warehouse introduced a lubrication service for its flat-cars and trolleys. Every week a number of them would be lubricated. Although they were numbered and service was duly recorded, there was no way of telling which of the trolleys had undergone the periodic service except by comparing their number on the list in the office, or turning them upside down. Since the trolleys were in constant use, some loaded and some empty, the lubrication team needed an easy way of identifying a free trolley that needed lubrication. To solve this problem a small metal plate painted green on one side and red on the other, was affixed in a slot on the side of the trolley. When a trolley had been lubricated the plates were turned to one specified colour until all had been completed. When the next cycle was started the plates were turned to the other side. Thus it was easily recognised which trolleys needed lubricating.

We now turn to the wide variety of electro-mechanical means of communication. By now internal telephones are slowly being superseded by the intercom, and with good reason. The instant response to "Are you there, Jim" is expedient, and the interlocking of groups calls and other switching advantages as well as paging, are very economical in time. In many instances paging is essential. However, excessive use of loudspeaker paging is not advisable. After a while it becomes either oppressive or inaudible. Paging by means of the pocket receiver is sometimes justified in spite of its cost. Closed-circuit television is becoming more widespread and for mobile crews over large areas, two-way radio is almost indispensable. Telewriters that transfer handwriting instantly to one or more stations are very advantageous. Messages can be received without necessitating anybody's presence next to the instrument. Leaving messages on tape, either from a direct recording or through an unanswered telephone call, can also be helpful. Such devices replace the unreliable services of some clerical help.

Thus we have at our disposal ways and means that can reduce the amount of time spent on communicating. Before adopting any of the more expensive solutions a survey of needs must be carried out, indicating the type of communication to be handled, its frequency, the number of stations needed and the distances involved. Investments in this area will prove to be very rewarding.

SETTING AND APPLYING JOB TIMES

The controversy surrounding this topic has been long, lively, far ranging and sometimes confused. Basic issues are often ignored and people tend to pursue their own subjective opinions. Let us then try to clarify some points, and start at the beginning.

Q. Are job times necessary for running an effective service?

A. No. Assuming we have stable conditions, right procedures, proper motivation at all levels and adequate supervision, a maintenance service can be effective without establishing job times, either before or after the job is done. Such a situation, however, is Utopian.

Q. If there are neither job times nor time-recording procedures *can* controls be exercised?

A. No. Control means the comparison of actual performance to pre-set objectives or limits and the computation of totals from details. Without job times we can neither exercise control over manpower requirements, nor quantify the work performed. It is also impossible to calculate the cost chargeable to centres or accounts.

Q. Is there a guarantee that job times *will* improve efficiency?

A. No. Job times, when available, facilitate the task of managing the department. It allows planning and allocating of jobs and may indicate where improvements are necessary. Job times tend to present workers with a motivational challenge which is presumed to have a beneficial effect. However, the introduction of job times is not always smooth and may raise more problems than it solves, mainly in the field of human relations.

Q. Is there any use applying job times *without* an incentive scheme?

A. Yes, most definitely! The assignment of jobs where times are allocated, contrasts sharply with a *laissez-faire* régime where everybody takes as much time as he pleases over any job. The allocation of job times forces the foreman to relate his thinking to the job in question and to the use of a full day for each worker. Similarly, the worker is encouraged to try his best to complete the job within the allocated time, not least because he may feel that he is being tightly controlled. Thus more work tends to be accomplished.

Q. Can *all* maintenance jobs be covered by job times?

A. If you mean whether you can have all of the times *in advance*, the answer is no. You proceed by first establishing times for oft-repeated jobs or jobs that vary within small limits. This may make up between 30 to 50 per cent of all jobs. Lubrication and inspection are examples of repetitive jobs for which ready-reference tables can be worked out. Most other jobs contain work elements that are repeated with variations and the total job time is then made up of these elements and others that are "guesstimated." Some jobs are truly imponderable but they add up to less than 15 per cent of total working hours (the "rusty bolt" argument).

Q. Is there any use having only *part* of the jobs covered by times?

A. Yes, certainly, since the exceptional hard-to-estimate jobs are in the minority.

Q. Is it advisable to spend so much effort on setting the times and administering them?

A. Yes. A collection of times is made progressively and the administration of job times once they are available should not constitute in excess of 5 per cent of the direct maintenance workforce, *i.e.* one estimator for twenty craftsmen. Would you allow twenty production workers to work without knowing how long jobs should take and how long they took in effect?

Q. Can job times be a substitute for the otherwise poor management of this function?

A. Definitely not! To expect this is one of the most frequent errors encountered. In the heat of an argument objectives are often forgotten or misunderstood. Job times are only a means to improve the implementation of the service. It can only be done *if other conditions are also satisfactory*. There must be a certain measure of planning, adequate shop conditions, an organised materials and spares store and good supervision. If these conditions are not met efforts to set job times will invariably fail.

When job times are discussed the following important points are to be considered:

Costs. In the early stages of a project establishing times and reference tables and administering them may require costly efforts.

Accuracy. Variability of elements due to changing work methods and conditions must be allowed for.

Coverage. A small proportion of work is imponderable. (There are some wild notions about this.)

Usefulness. Applications and uses of job times should be foreseen and maximised.

The benefits of applying job times on a wide range of jobs, can be summarised as follows:

(*a*) Increased performance of the workforce. *Even when times are not used to pay incentives,* output of the workforce increases sharply. Rises in effective utilisation may follow a pattern which increases the lower the starting-point, as shown in the following table of approximate figures showing a gradual increase in manpower effectiveness.

	%	%	%	%	%
When present effectiveness is:	35	45	55	65	70
Expected utilisation can reach:	60	70	75	80	85
Therefore increases performance by:	42	36	27	19	17

Some of these figures are supported by published case material. Achievement of such results will be due to the general sense of purposeful activity generated in the department by the introduction of job times and not necessarily as a result of bonuses.

(*b*) Improved scheduling, balancing and manning of teamwork. A reduction of accumulated work will also result from assigning several jobs to each worker at the start of the day, and from the reduction of waiting time between jobs.

(*c*) Supervision becomes more effective since there will be tighter planning. Instead of time spent on allocating craftsmen to jobs or repeated reassignments during the day, more time can be spent on actual supervision and training.

(*d*) Time summaries will allow assessment and recording of workers' efficiency and departmental performance.

(*e*) Jobs consuming excessive time are highlighted and the continued use of machines which may be too costly to maintain can be investigated.

(*f*) The need for training can be identified where necessary.

Methods for accumulating job times vary over a wide range in their respective accuracy and the time required for collecting them and the overhead costs incurred in their introduction will vary accordingly.

The sequence of methods in ascending order of accuracy and costs is approximately:

"Guesstimating."
Historical data or past performance.
Estimating.
Analytical estimating, job-slotting.
Work sampling.
Time studies and production studies.
Universal Maintenance Standards (UMS).[36, 44]
Synthetic time data and Predetermined Motion Time Standards (PMTS).

The first two must be dismissed as too inconsistent and unreliable. Estimating depends too much on the estimator. Analytical estimating, however, requires the job to be broken down into elements which can be more accurately identified and plotted. Some elements can be compiled from time studies and a small percentage, not more than 20 per cent, can be estimated. Results of performance on the basis of these times must be recorded so as to learn from experience, and readjusted if excessive discrepancies occur between estimated time and actual performance. Work sampling is discussed below.

The undisputed championship of job times in maintenance goes to Imperial Chemical Industries (ICI) which pioneered the collection of data into systematic reference tables. What has been accomplished there under the leadership of the late R. M. Currie is most impressive. Since maintenance is probably the major cost component in the erection, running and maintaining of a chemical plant, the company found it economically feasible to method study a great number of jobs.

A direct extension of methods improvements is timing or, more precisely, work measurement. Since work could be measured in a variety of ways according to length of job, repetitiveness, manning and variability many different methods were employed.

Traditional time-study is an expensive technique which can seldom be justified in maintenance. It requires a much too high degree of repetitiveness of jobs to be studied and they must be reasonably similar in content. It can therefore be used only on fractions of larger jobs and the results can only serve to support analytical estimating techniques. Elements that frequently recur—such as greasing, cleaning of filters, replacing of lamps, caulking of boiler tubes and a host of work with nuts, bolts and screws—can be profitably time-studied, provided one's scale of operations is large enough to justify the employment of a time-study man. Because of the large scale of ICI's operations, job times could be applied under varying conditions and frequently rechecked. The resulting times were all incorporated into tables and charted. At present there are few jobs at ICI for which times cannot be set on the basis of available data.

Now, to be sure, it must be realised that few companies can imitate ICI. A few large concerns in the U.S.A., such as DuPont, have pursued similar activities for the last two decades. Since processes change constantly and plant expands and the installation becomes more complicated, this job is never finally accomplished. To attempt complete coverage, therefore, is not feasible. Smaller plants can adapt the method to their needs by classifying jobs, recording performances and using bench marks. This is often called "job-slotting." By relaxing the demand for 100 per cent coverage and lowering expected accuracy, a store of data can be built up. This data is supplemented by using the know-how of craft supervisors, and the estimator can then ensure a coverage of approximately 85 to 90 per cent. It has been estimated that one time-analyst can

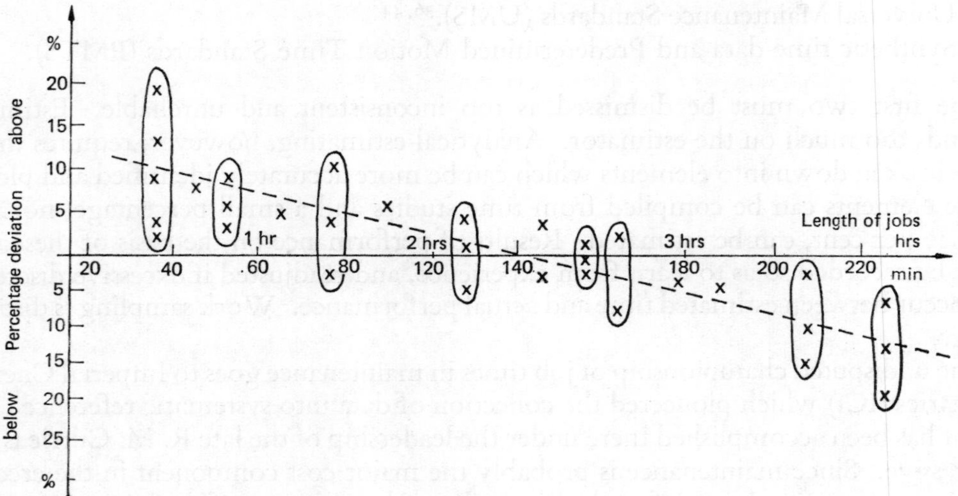

Fig. 64. Control chart for job-time estimates. The deviations from estimates are plotted above or below the horizontal time-scale.

provide job times for about twenty to twenty-five craft workers. When control curves are plotted, the accuracy of estimates can be expected to increase over a period of time. The deviation of actual work from the estimate is to be plotted. Per cent value should be plotted over or below a horizontal co-ordinate as shown in Fig. 64.

Several systems which allow a "synthetic" compilation of times from reference tables have achieved a measure of acceptance. MTM and Work Factor are being used successfully in production work; however, their application to maintenance is less than popular. They are costly to apply and the main difficulty lies in the fact that the estimator must have a very good knowledge of what every job entails *before* starting to work on his estimates. This is often impracticable and very time-consuming.

A sensible approach called "Universal Maintenance Standards" was introduced in the mid 1950s by the developers of MTM: Messrs. Maynard, Inc. of Pittsburgh, U.S.A. The method relies on the fact that most jobs are not performed under standard or identical conditions. Therefore, times vary within a certain range and a single allowed time derived from the past cannot be applied on all occasions. In UMS practice, however, jobs are classified into groups tabled along a time range and "benchmark" jobs are written along the time-scale. The job times themselves are calculated from past records by a number of formulae to ensure reliability and range of variability. When a job is estimated, reference cards are consulted and the total time is the summary of a number of elements identified and extracted from the cards. The source for each element is recorded on the estimate sheet. In order to reconcile each element with the actual prevailing conditions, the work to be done is surveyed and craftsmen are consulted. An example of a UMS table appears in Appendix III.

To satisfy promoters of the "rusty bolt" argument, the tables take into account the condition of components, accessibility at the workplace and other variables which may effect working time.

The adoption of ready-made data from outside sources must be treated with caution. Times are not usually repeatable in different locations. It is very seldom that one really understands the way these tables were compiled and under what assumptions they were expected to be applied. In other words, do they include allowances? Are they averaged performance times or normalised values, etc? The most difficult aspects to compare are work methods, tooling, sequences and surrounding working conditions. Sometimes the skills of craftsmen are on a different level too, not to mention motivation and supervision. Thus "foreign" data must not be taken at their face value.

To illustrate this point the example of car repairs can be cited. Ford garages throughout the world are provided with tables for standard jobs such as clutch adjustment, cleaning the carburettor, etc. Expected performance times are tabled for models and vintage. The tables also provide for variability when certain jobs are performed in conjunction with other jobs, such as the dismantling of a gearbox or the removal of an air filter for cleaning. When performed in sequence, the composite values in such cases are *not* the sum of the individual jobs.

When it comes to the application of these times, in garages round the world conditions usually differ from "averages" assumed or assessed in the place of origin. Variations in space and available facilities can, and often do, double performance time. When this effect is studied it is usually found feasible to use a correction factor which will bring the values in line with prevailing conditions. Most garages will then use the tables as guidelines for allocating jobs. Unfortunately, the customer usually pays on actual performance whatever its level and he is not billed by the value that appears in the table.

Procedures for time-setting by estimates are similar to those used in computations based on time studies. Normal times for elements vary according to work content and can be read off from charts and tables. An example relating to welding appears in Fig. 65. Fixed elements are added to cover the necessary preparations, such as getting instructions, tools and materials, cleaning up and returning to base. It is advisable to use specially designed forms (*see* Fig. 66) that guide the estimator and ensure that all elements are included. It will also allow the estimator to refer back to a previous job

ARC-WELDING—HANDLING TIMES IN MINUTES AT 133% RATING WITHOUT ALLOWANCES	Section	Number
	63	1/12
	Issue no.	2
	Date	March 1970

Set-up and tear down

Receive work—study—bring M/C—prepare workplace
Return tools—material—Equip.—clean W/P—check out
Obtain fixture—tools—electrodes from store
Set up or manufacture fixture

Handling time/piece

Load and unload in position to weld — Part wt. kg.	0·50	2	15	30
In assembly—forced fit	0·18	0·27	0·43	0·78
On table	0·10	0·14	0·24	0·45
In vee blocks/parallels	0·13	0·18	0·30	0·55
In fixture or vice	0·12	0·17	0·28	0·50
Against angle plate	0·15	0·23	0·36	0·66
Turn piece over or turn 90°				
On table	0·04	0·06	0·11	0·20
In vee blocks or parallels	0·06	0·08	0·14	0·26
In fixture or vice	0·05	0·07	0·12	0·22
Against angle plate	0·08	0·12	0·19	0·35
Gauge or line-up part				
Scale	0·12	0·16	0·21	0·36
Square	0·16	0·23	0·36	0·45
Level	0·10	0·13	0·16	0·20
To hole or line up by sight	0·07	0·10	0·12	0·15

Tighten and release holding devices

T-bolt—stud—clamp	1·50	Strap clamp large	1·70
In vice by hand	0·12	In vice with soft hammer	0·32
C-clamp (up to 200)	0·63	C-clamp (over 200)	2·00
Toe or vee clamp	0·80	Set screw	0·38

Clean and brush

Vice	0·19	Vice with parallels	0·36
Small table	0·17	Large table	1·70
Small fixture	0·23	Large fixture	2·30
Blow-off surface or hole	0·05	Wipe surface—per sq. m	0·30

Shim and align part

Length part in mm	Width (mm) 250	500	750	1000	1250	1500
—120	1·00	1·75	2·50	3·50	4·50	5·25
—180	1·50	2·50	3·50	4·75	6·00	7·25
—240	2·00	3·50	5·25	7·00	8·75	10·25
—300	3·00	5·00	7·25	9·50	10·00	14·50
—400	4·50	7·50	11·00	14·50	18·50	22·00
—550	6·00	10·00	15·00	20·00	25·00	30·00
—600	7·50	13·00	19·00	25·00	39·00	38·00

Fig. 65. Examples of standard times for estimating manual welding.

Date: 5/11/71 *Sheet:* 2 *of* 3 *Job no:* 23476 *Est. no:* 1084

Part: Tank base *Oper. no:* 2, *Dept.* Fabrication *Quantity:* 2

Operation: Weld plates as shown in sketch-plates, supplement, fitted up

Material: 8mm. mild steel plate *Labour req:* Welder *Estimator:* v.p.r.

Set up and tear down			Freq.	Min.	Total
Receive work—study—bring M/C—prepare w/place			1	10·0	10·0
Return tools—material—equip.—clean w/place—check out			1	5·0	5·0
Obtain fixture—tools—electrodes from store					
Set up or manufacture fixture			1	2·0	2·0
					17·0

1st pass Angle/butt/vee-groove Body/root/back
Electrode: univ. 4 Current: 150 A Length-pass (cm): 300 mm
Equiv. "W" mm: 4·5 Position: Flat Access: Easy

Handling per part			Freq.	Min.	Total
Electr. handling for start and complete weld		0·06	4	0·06	0·24
Electr. handling for intermittent welding		0·10	2	0·10	0·20
Tack welding		0·15/0·30	12	0·15	1·80
Weld	50 length pass (cm) x weld min/cm	0.03	10	1·50	15·00
Change electrode	length pass (cm) x $\frac{0{\cdot}09/0{\cdot}15}{\text{cm/ electr.}}$		10	0·09	0·90
Move and adj. tools	length pass (cm) x 0·003/0·005		30	0·003	0·03
De-slag and or clean	length pass (cm) x 0·03/0·05/0·015		100	0·015	1·50

2nd pass Angle/butt/vee-groove Body/root/back
Electrode: Univ. 5 Current: 225 A Length pass (cm): 300 mm
Equiv. "W" (mm): 4·2 Position: Flat Access: Easy

Handling per part			Freq.	Min.	Total
Electr. handling for start and complete weld		0·06			
Electr. handling for intermittent welding		0·10			
Tack welding		0·15/0·30			
Weld	length pass (cm) x weld min/cm.				
Change electrode	length pass (cm) x $\frac{0{\cdot}09/0{\cdot}05}{\text{cms/ electr.}}$				
Move and adj. tools	length pass (cm) x 0·003/0·005				
De-stag and clean	length pass (cm) x 0·03/0·05/0·015				

Notes or sketch			
	Minutes per piece		19·67
	Minutes per batch of 2 units		39·34
	Min at 133 per set-up		17·00
	Min at 133 per order		56·34
	Plus 3% U/delays + 20% pers. and fat.	12.80	69·14
	Plus 33% incent. allow. (standard min)		92·00

Fig. 66. An example of an estimate sheet for welding work, giving all relevant factors.

when necessary. In case of arguments over time allowed, the form serves as a basis for discussion.

It is probable that from the experience gained during work, many readers will be able to adapt a solution that will best suit his needs. One fact is indisputable: *time and effort must be spent on this task* and no matter how diligently we work *it is never completed.* However, in view of the potential benefits the expense is well worth while.

Earlier remarks must here be emphasised: Job times are useful for many purposes *other* than the payment of incentives. The latter should perhaps be considered as an eventual by-product. The uses of job times for scheduling, costing and management data are important enough in themselves. In fact, the use of job times for incentive detracts from their other uses and often introduces pressure which diminish their accuracy. Discussion should shift from "Why do job times make incentives so complicated?" to "Why can't we find more expedient methods to pay incentives?"

It remains for us to discuss the administrative steps involved in applying job times. The time required to perform routine repetitive jobs is recorded on the service schedules: these are allowed hours. Non-repetitive jobs are assessed individually by the estimator and recorded on job cards or individual workers' time cards. Emergency jobs cannot be estimated before the work is performed and retroactive estimates are required.

To allow effective use of accumulated data, quick-reference tables must be laid out so as to minimise computation time. A special technique called "multi-variable charts" has been devised by Phil Carroll.[4]

The physical shape of reference tables is also of importance. The use of visible card indexes is advocated while the use of ordinary folders or binders is not recommended. Ring binders supported at an angle of about thirty degrees are commercially available. Pages can be laminated in acetate to withstand everyday wear and tear.

ACTIVITY SAMPLING APPLICATIONS (WORK SAMPLING)

In its early stages activity sampling used to be called "ratio delay study." It was intended to be used for separating the percentage of lost or ineffective time from productive time and as a technique for the calculation of production output. Percentages of unavoidable elements resulting from these studies were added in the form of allowances to make up the total allowed time.

In the twenty-five to thirty years since this technique was first introduced, no other has arrived on the scene to detract from its original value. Activity sampling has blossomed into a fully-fledged and well-documented practice and has been adapted to the latest trend in computerisation. For example, the coding observations can be transferred directly on to punch cards by means of magnetic markings or a portable punching device *e.g.* IBM's "Port-a-punch."

The principles of work sampling are well rooted in statistical theory. The percentage of a chosen element can be established very accurately and by now alignment charts and tables of per cent deviations have streamlined calculations to the barest minimum.

Since the actual practice of activity sampling is well covered in the literature,[23] we will limit our remarks to discussing its applicability to maintenance.

In what ways can this valuable tool assist us? Although in certain applications it can provide information on job times, this is not its best application. The areas to which work sampling is particularly well suited are:

(*a*) Distribution and frequency elements, *i.e.* how much of certain elements is done by each member of a team of workers.
(*b*) Manpower utilisation, *i.e.* identifying the percentages of effective and ineffective elements.
(*c*) Establishing frequency of occurrence, *i.e.* workers waiting for tools, machines operating or idle, etc.
(*d*) Effort rating surveys.

One advantage of this technique is the fact that it requires no outside consultants, and should be carried out by local staff who have undergone adequate training. Although it is claimed that a short demonstration in the use of this technique is enough to teach most people how to apply it, the value of a well-trained analyst is not to be lightly dismissed.

While the actual procedure is simple, there are a number of steps that must be observed when a study is made.

(*a*) People concerned must be well informed about the purposes of the study. (An observer appearing at odd times may cause some uncomfortable reactions.)
(*b*) Workers and their representatives should be consulted and advised.
(*c*) The intended use of results should be defined in advance.
(*d*) Precautions should be taken to avoid biased results by taking enough random readings at well-planned intervals.

The form shown here (*see* Fig. 67) provides space for marking the alternative elements by letter-coding, *e.g.* *w*–working, *i*–idle, *t*–transport, etc. Each reading, however, could be further specified by additional codes. Thus the figures 1 to 5 could be added to a "*w*" reading to express effort rating between "poor" and "excellent" or otherwise normal percentage figures may be used. The letters "X" and "O" could be used to indicate necessary or unnecessary transport.

This versatile technique can also be easily adopted for checking the work of craft teams (Fig. 68). The condition of equipment can also be checked by ticking off conditions specified in a checklist and the incidence of readings will allow the percentage assessment to be made.

Another use of work sampling is in the support of incentive plans. Certain incentive schemes require either periodic sampling to provide over-all rating factors, or alternatively to specify levels of activity at which bonuses will be paid. This information can be conveniently provided by work sampling.

WORK SAMPLING SHEET	Study no.
Dept/Shop/Team..	Serial no.
Observer.. Sec........................	Date

Obs. / Random times	Operator/Machine/Process 1	2	3	4	5	6	7	8	9	10	Tally count — Codes and observations *w*	*t*	*a*	*n*	*p*	*i*	*x*	No. of obs.
1																		
2																		
3																		
4																		
5																		
6																		
7																		
8																		
9																		
10																		
11																		
12																		
13																		
14																		
15																		
16																		
17																		
18																		
19																		
20																		
21																		
22																		
23																		
24																		
										Total observations								
										Percentages								100%

Codes:

w—working
t—transport
a—absent from workplace
n—writing, recording
p — personal
x — auxiliary
i idle, waiting

Fig. 67. Work-sampling study sheet. The observer makes a tour of operators or machines according to the random lines marked along the scale (10–24 tours per day). Each observation is recorded in code letters and later summarised; accuracy of calculation increases with the number of readings.

CRAFT-TEAM ACTIVITY SAMPLING

Date:			Mechanical team						Electr.	
Time	Jobs and work location / Names of craftsmen		Williams	Johnson	Blakeman	Peterson	Powell	Rogers	Walker	Mannings
8.20	Powerhouse	R	105	115						
			W	W						
8.40	Building no 5	R			–					
					A					
9.00	Frame and body assembly line	R				90	110			
						W	T			
9.20	Machine no. 32	R						85		
								W		
10.00	Air conditioners	R							105	95
10.20									W	W

Fig. 68. A craft-team activity sampling sheet (R = rating, here 100 per cent = normal).

BREAKDOWN ANALYSIS

The recorded plant history of machines contains useful information apart from the obvious indication of its past repairs and incurred costs. When this information is analysed in the right way it can disclose:

(*a*) inherent weaknesses of design;
(*b*) poor utilisation practices (euphemism for "abuse");
(*c*) inadequate maintenance;
(*d*) unsuitability of replacement parts;
(*e*) typical sequence of failures.

More insight into the reasons for breakdowns can be gained when several identical units are in operation and the breakdowns are tabled in parallel. In some instances collection of data by section or department may show areas where more frequent service could reduce the occurrence of breakdowns. Statistically the results can be made more significant by surveying a wider "population" or by covering longer periods.

The latter alternative has a limited use since over a long period the condition of a machine may deteriorate considerably, and then, of course, we are no longer dealing with the same machine.

A reliable source of information for the required data is the work order. The cause of a breakdown should appear on the work order as a matter of routine. The date of a breakdown as well as the completion of its repair must also be recorded.

The cost of every breakdown repair will be calculated by the accounting department on the basis of the hours worked, materials issued and overhead and the total is recorded on the job card. When the job cards are returned from accounting to maintenance for filing, the maintenance clerk records the summary of details on the equipment history card. At the same time it is advisable to post the event on a log sheet. This will facilitate further analysis and obviate the need to leaf through hundreds of

BREAKDOWN LOG BOOK							
Date of event	Plant unit	Work order no.	Dept.	Cause of breakdown	Total downtime	Total cost of repair	Remarks action taken

Fig. 69. Log-book for recording breakdowns chronologically.

completed work orders (*see* Fig. 69) when we eventually decide that an analysis of breakdowns is due.

In fact, if analysis of breakdowns is to follow the usual routine, the use of a "keysort" card as a work order is justified. This card has a code printed along its margins and it is punched according to the code (*see* Fig. 13). Classification is achieved by passing a "skewer" through the code holes and lifting the bunch of cards. Those remaining in the rack are the ones pertaining to that group. By consecutive sorting the cards are grouped. Thus, for example, we pass the skewer through "breakdowns" and all job cards relating to breakdowns remain in the rack. Next we take this group and pass the skewer through "pumps" and we are left with cards relating to pump breakdowns. Subsequent sorting will group the causes one after the other. When that is done the breakdown analysis sheet can be completed (Fig. 70). By means of this form we can detect certain trends in the frequency, and severity of breakdowns and their costs. Excessive costs of breakdowns could indicate that it would be preferable to increase expenditure on maintenance services so as to avoid some of the breakdowns. The frequencies of breakdowns due to the different causes will indicate where corrective action is required.

Plant history for a single unit can also be profitably investigated. Here it is important to watch for a recurring fault and its frequency for a sequence that is repeated over a certain period. For this purpose, the breakdown record sheet is useful (*see* Fig. 71). It

ANALYSIS OF BREAKDOWN for the year 1971							
	Cause of breakdown		*Frequency*	*Severity stoppage in hours*		*Cost of repairs (£)*	
	Detail	*Code*	*Number of occurrences on all units during the period*	*Total hours lost*	*Average severity of breakdown*	*Total cost of repairs*	*Average cost of repairs*
Group A Pumps	Normal wear and tear	N	5	15	3	300	60
	Lack of maintenance	L	2	20	10	100	50
	Improper use	I	3	18	6	450	150
	Accidents	A	1	10	10	500	500
	Faulty design detail	F	2	3	15	100	50
	Other reasons	O	5	10	2	60	12
	Total		18	76	4,2	1510	84
Group B Mixers	Normal wear and tear	N					
	Lack of maintenance	L					
	Improper use	I					
	Accidents	A					
	Faulty design detail	F					
	Other reasons	O					
Totals	Normal wear and tear	N					
	Lack of maintenance	L					
	Improper use	I					
	Accidents	A					
	Faulty design detail	F					
	Other reasons	O					
	Total						

Fig. 70. Breakdown analysis sheet. The above shows an annual summary for detecting trends in the frequency and severity of breakdowns.

is suggested that this type of investigation be carried out on critical units only and for specified periods. The procedure could otherwise prove very costly.

The tabulation of results either statistically or chronologically can reveal:

(*a*) The need for training of machine operators.
(*b*) The need for rescheduling of services.
(*c*) A gap in responsibility or an oversight in service instructions.
(*d*) An unsuitable component.
(*e*) A fault in design, or installation, *e.g.* misalignment.
(*f*) Poor maintenance workmanship.
(*g*) Sensitive parts of the equipment.
(*h*) The need to replace a particular machine.

BREAKDOWN RECORD SHEET

Plant equipment, name and description: | *Plant no.* | *Location:* | *Period covered:*

	Dates / *Parts affected*	*Jan 12*	*Feb 12*	*May 3*	*July 19*					*Repeated events*
Mechanical	Bearings	X								
	Shaft		X							
	Housing	X								
	Reduction gear									
	Power transmission		X							
	etc.									
Electrical	Starter	X								
	Motor				X					
	Brushes				X					
	Instruments		X							
	etc.									

Fig. 71. A breakdown record sheet for one machine.

The experience from a plywood factory could here be cited to show that often maintenance people will accept a situation without questioning the reasons.

In this particular factory, electric motors were burning out at an excessive rate. For some unidentified reason the machining parts driven by the motor seized too frequently. While the operators tried to free the blades the motors burnt out. Thus the electrical team was kept busy replacing and rewinding motors. A short survey showed that the exhausting fans for chip and sawdust were in proper working order but that there was excessive accumulation of sawdust in the ducts. Further investigation showed that the break-switches activating the exhausters were next to the main switchboard, while the production machines had their individual "start–stop" controls in the usual way.

Upon enquiry it transpired that responsibility for activating the exhausters was not fixed on anybody in particular and for this reason they were not always started *before* the production machines. When an amount of sawdust collected that the exhauster could not dislodge, it continued piling up until the unavoidable happened.

To remedy the situation a green light was then installed in a prominent place, and instructions were issued not to operate machines until the light was on indicating that the exhausters were in operation. As a result the load on the electrical repair team decreased dramatically.

Breakdown analysis cannot be complete without discussing the way in which the results should be presented. Time lost due to breakdown should be shown as:

$$\frac{\text{Total downtime due to breakdowns (Dept. A. period X)}}{\text{Total units produced (units, tons, yards, etc.)}} = \text{downtime per unit of production}$$

Another control that indicates whether breakdowns are running wild or not, is the percentage of work that they generate, out of the total clocked maintenance man-hours:

$$\frac{\text{Total hours spent on repair of breakdowns} \times 100}{\text{Total of actual maintenance hours paid for}} = \text{percentage of breakdown work}$$

When these two indices are charted on a monthly basis a general trend will emerge. In a broad sense it will show how well controlled the operation of equipment is.

Chapter 9

The human element

The observation "The human element in maintenance has been sadly neglected" is by now a sad cliché. Assume for a moment that you have found a factory manager who is genuinely concerned for the well-being of his employees—who does he worry about?

A research project is hardly necessary to find the answer. First, the factory manager worries about his production workers, then about his sales force, then come his supervisors and finally his maintenance workers. With this order of priorities the state of low morale obtaining in the maintenance department is easily understood and it also helps to explain the accusations levelled against members of the department, such as low level of performance, lack of interest and frequent demands for higher pay.

Although it is not always easily apparent, nevertheless it is true that the *attitude of maintenance staff merely reflects the attitude of management.* In order to break this vicious circle management must adopt a positive attitude towards its maintenance workers, it must seek to understand their problems. It would be hard to assert that anything in human nature has changed since the Hawthorne experiments (at Western Electric in 1927) which proved that the interest shown by management towards its workers is a powerful motivating force for good. Few people familiar with maintenance work will dispute the fact that many of the particular needs of maintenance workers have been ignored for too long.

In this chapter the main influences affecting the human element are analysed. Efficient procedures and techniques can contribute a great deal to good performance. If our full manpower potential is not being used we are wasting a part of our resources of skill and time. This may add up to a considerable amount. If figures were available the following formula would apply:

$$\text{Manpower Effectiveness} = \text{Human Potential} \times \text{Application Factor}$$

The Application Factor is the product of Morale and Management Factors and since they are never both 100 per cent perfect, the Application Factor always has a value of less than one.

$$\text{Application Factor} = \text{Morale} \times \text{Management Factors} < 1{\cdot}0$$

The effectiveness of manpower lies in that part of the human capacity which is applied during work. It depends on the potential that is available and on the way it is applied. The factor of application is always less than 100 per cent, since neither morale nor managerial effectiveness can ever reach 100 per cent—there is always something left to be done. On the other hand, if at the start we assume the available Human Potential to have a value of 0·8 per cent it can be increased by developing it. Thus over a certain period if we were to keep up our efforts to improve these factors, we would observe a gradual rise in manpower effectiveness. This calculation is hypothetical since a way to quantify these factors has not yet been established. However, the formulae highlight the possible means of improvement.

"WHAT MAKES HIM APPLY HIMSELF TO THE JOB?"

The industrial worker spends a large part of his time at work. For production people it is a five-day week, but for maintenance staff it is often a six- or seven-day week and with overtime added at odd intervals. Production workers can usually watch their output rise and their performance can be recorded. At the same time they are remunerated for visible and tangible results. They can derive satisfaction from work detached from human relationships if they so choose. Apart from occasional contact with a quality inspector or with shop supervisors they can perform their jobs without noticing deficiencies in the organisation and being irritated by them. The same does not hold for the maintenance worker. He may be as keen as anybody to perform well, only to find himself hindered by bad working conditions, *e.g.* lack of spares, poor shop organisation or the constant misuse of machines. He gets upset not because he is more susceptible, but rather because he is in contact with factors that give rise to frustration.

A short survey of any maintenance team will reveal attitudes of frustration, despondency and lack of satisfaction. The following expressions are often heard:

"There's nobody to talk to. They simply won't listen."
"It's a dead-end job, no use trying!"
"Why hurry? The stores/parts/tools/foreman will hold you up anyhow!"
"It's a dirty job, they blame you for everything."

In fact, what are the necessary incentives and working conditions which will persuade a man to do his best? We show concern for morale when it is noticeably low. When things go smoothly nobody worries, it is as if "morale" were non-existent. However, morale is present whether good, mediocre or bad but we are not always aware of it. Yet the "morale climate" is important because ultimately it is responsible for good, medium or poor performance.[30]

Open bickering and grousing are not the only symptoms which point to a low state of morale. A quiet passive attitude, an indifference to work, are equally indicative of low morale and as harmful to the company as open criticism might be. The performance of a sullen workforce is no better than that of a rebellious one. Thus when we examine the prevailing morale, not only visible outward signs should serve us as indicators.

What is required, is a willing, active team that contributes not only its presence and part of its muscle-power, but also creatively participates in work. What is really needed is that little bit of extra care in aligning the bearings with an occasional suggestion as to how to solve a certain problem. To achieve this it is not enough to pay the going wages, to provide the agreed benefits and to apply a certain measure of supervision. Wages and benefits are soon taken for granted and the supervisor accepted as part of the familiar scenery.[37] On the other hand, what is all too real for almost every moment of working time, are the *prevailing conditions of the surroundings, the facilities in the shop*, the *tools in use* and *the service at the stores*. The personal *relationships between team-mates*, the attitude adopted by workers in other departments and the *indirect feeling of interest or neglect shown by management*, these are all part of the real climate surrounding a maintenance worker. Throughout his hours of work a maintenance worker is most conscious of these factors.

What then are the controllable factors that help in creating the proper frame of mind called "good morale?" The required ingredients originate from:

MANAGEMENT POLICIES—which relate directly or indirectly to maintenance,
WORKING CONDITIONS AND RELATIONSHIPS—under which work is performed, and
PERSONNEL PRACTICES—which are in force.

Undoubtedly, there are other factors that affect the individual's morale, such as personal hardships, preferences and dislikes and many outside influences over which management has no control. Discordant attitudes and relationships between people may develop and these cannot be ignored, although their effect can be minimised in an otherwise good atmosphere.

The answer lies with management and in the multitude of factors that lead to good performance (*see* Fig. 72). Those measures that we know and apply in the case of production workers are doubly important in relation to maintenance.

The combined effect created by these factors acts as *a motivator or as a deterrent* to the input of the available manpower potential. It is comparable to the effect of a charged grid in an old-fashioned vacuum tube. The more the grid is charged, the less positive particles can pass through. The maintenance man is confronted with an atmosphere (a grid) that is more or less conducive to his positive contribution to work.

Few maintenance managers will dispute the fact that little attention is devoted to the procedures that motivate the maintenance worker. In fact little attention is paid to maintenance at all. Among other things:

(*a*) Very seldom are maintenance people encouraged to speak about their problems at work. Their impressions, opinions and suggestions are seldom sought.
(*b*) The organisation structure is almost never reviewed, maintenance often grows wild like a parasite on a tropical tree.
(*c*) Neither the team nor the maintenance supervisor can win a favourable hearing for their complaints.

In the end, this neglect permeates through the ranks and poisons the atmosphere.

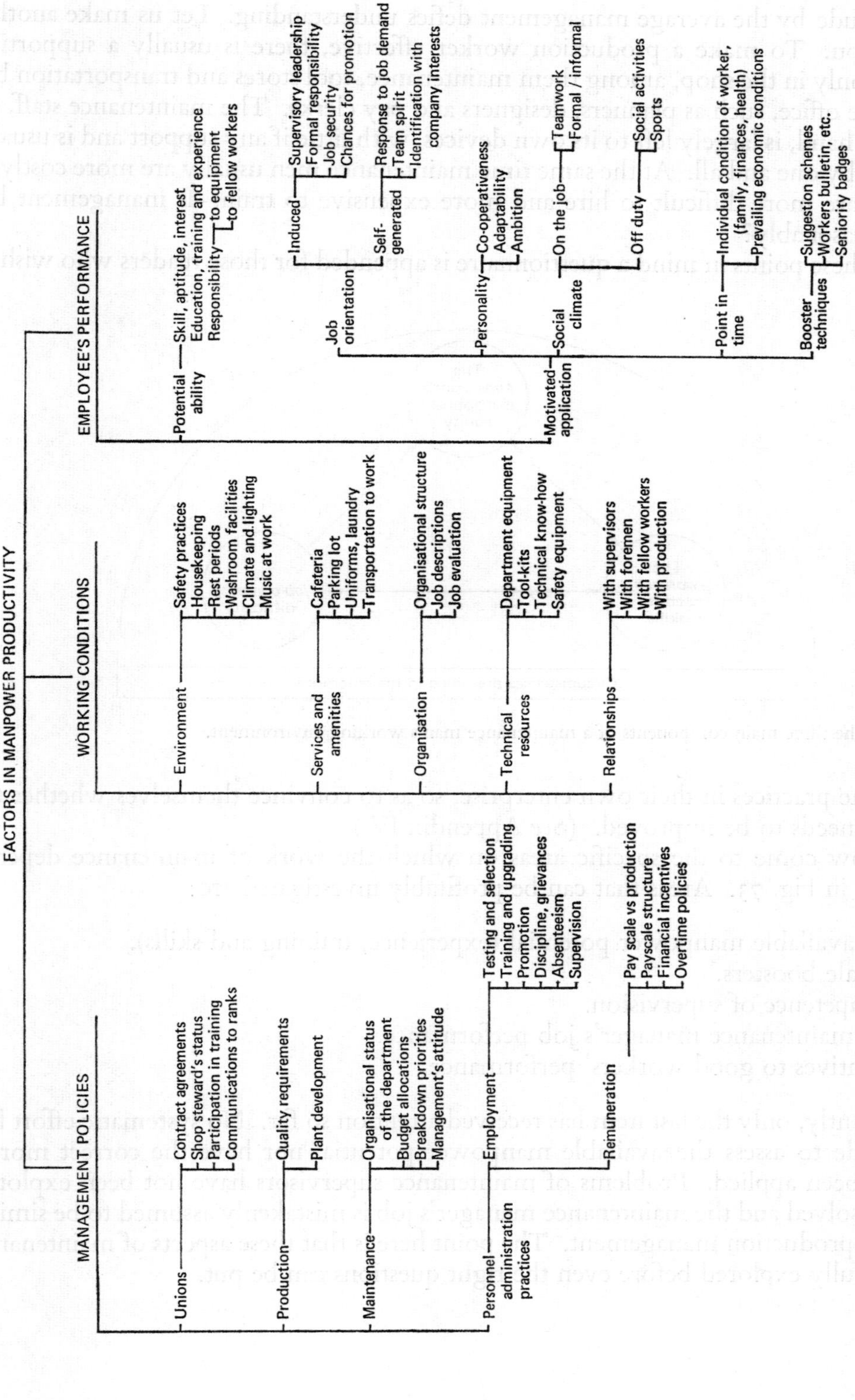

Fig. 72. The grouping of relevant factors affecting manpower productivity in maintenance.

This attitude by the average management defies understanding. Let us make another comparison. To make a production worker effective, there is usually a supporting staff not only in the shop, among them maintenance, tool stores and transportation but also in the office, such as planners, designers and pay clerks. The maintenance staff, on the other hand, is largely left to its own devices, with little if any support and is usually criticised by one and all. At the same time maintenance men usually are more costly to employ, are more difficult to hire and more expensive to train. Is management behaving reasonably?

With these points in mind a questionnaire is appended for those readers who wish to

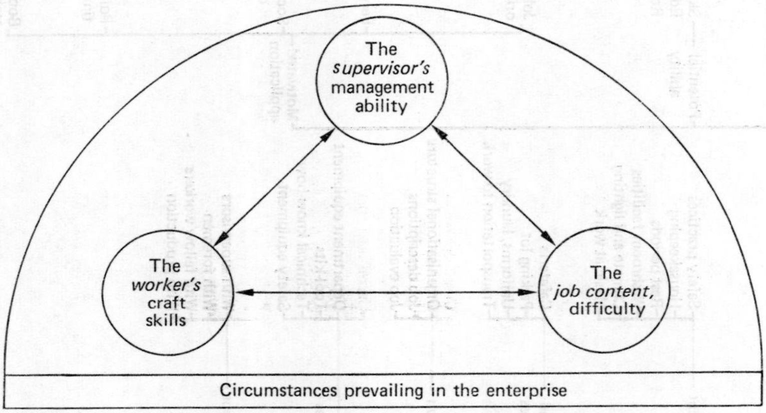

Fig. 73. The three main components of a maintenance man's working environment.

review the practices in their own enterprise, so as to convince themselves whether the situation needs to be improved. (*See* Appendix IV.)

We now come to the specific areas on which the work of maintenance depends as shown in Fig. 73. Areas that can be profitably investigated are:

The available manpower potential (experience, training and skills).
Morale boosters.
Competence of supervision.
The maintenance manager's job performance.
Incentives to good workers' performance.

Apparently, only the last item has received attention so far. No systematic effort has been made to assess the available manpower potential nor have the correct morale boosters been applied. Problems of maintenance supervisors have not been explored let alone solved and the maintenance manager's job is mistakenly assumed to be similar to one in production management. The point here is that these aspects of maintenance must be fully explored before even the right questions can be put.

MAINTENANCE MANPOWER POTENTIAL

The question arises whether management knows the potential of the maintenance manpower at its disposal. Too often management's impressions are based on conjecture and the senior executive in maintenance is only vaguely aware of "what he has or what he hasn't." When maintenance staff are originally recruited they pass through the regular screening of a personnel manager. This is the only screening that quite a few members of the team ever receive. When transfers are made from one department to another no screenings take place and the new member of a team is assumed to be competent unless proven otherwise. If his abilities were low at the start, he is expected to improve by his own efforts.

Since often no personnel records are kept in the maintenance department, the individual worker's abilities, skills, special talents are known only to his supervisor, and even then only vaguely. In fact more often than not, the total available potential is unknown. With this deficiency in mind a review of manpower potential is shown in Fig. 74.

The fact that the characteristics of a good maintenance man differ from those of a production man has only recently been recognised. What, then, makes a good maintenance worker?

He must be inquisitive and resourceful.
He must be adaptable and self-sufficient.
He must be dependable and responsible.
He must be keen to learn and interested in his trade.
He must be able to observe and visualise.
He must be co-operative and helpful.
He must be healthy in body.
He must be emotionally stable.
He must be safety conscious for himself and his fellow workers.
He must be able to work under pressure.
He must be alert and possess keen senses.
He must always try to do a good job.

Some of these traits are measurable indirectly by questionnaires. Tests have been devised for this purpose and they should be applied during recruiting procedures. Some of the traits can be developed through adequate training.

When a worker applies for a production job he expects his position and his chances for promotion to be clearly defined and he expects to be promoted if he proves his worth. The situation prevailing in maintenance is not at all the same. Nine out of ten mechanics or electricians will believe that theirs is a dead-end job since there is little obvious scope for advancement. To dispel their misgivings a promotion policy should be established and given adequate publicity. Promotion in terms of seniority, responsibility and status should all be included in this policy.

It has often been observed that members of the maintenance staff remain in the same

QUESTIONNAIRE

The available manpower potential

Rate the maintenance team and give breakdown in points up to 100 the following personnel aspects:

(The total in each line should be 100 points)

		Rating points		
		High	Medium	Low
1.	Formal technical education			
2.	On-the-job training completed			
3.	Experience on the job			
4.	Knowledge of the job			
5.	Craftsmanship at work			
6.	Trade skills and dexterity available			
7.	Responsibility shown to property			
8.	Responsibility shown to co-workers			
9.	Responsibility shown towards own job			
10.	Dependability			
11.	Perseverance in efforts			
12.	Temper under control			
13.	Personal self-confidence			
14.	Initiative and judgement			
15.	Creative thinking and curiosity			
16.	Alertness of senses, observation and memory			
17.	Deductive thinking			
18.	Teamwork and co-operation			
19.	Performance of unsupervised work			
20.	Taking of orders			
	TOTALS			
	Divide by 2,000 to get overall %			

Fig. 74. A specimen list of questions for surveying the available maintenance manpower potential.

job for many years with supervisors and foremen especially having many years of seniority. The trouble is that these people practise their trade *as they originally learned it* many years ago. Upgrading exists sometimes but updating is very rare. Since maintenance people usually learn by doing, it is expected that they will continue in this way. The snag is that nowadays technical progress is advancing at a faster speed than most people can keep pace with, however hard they try, and pretty soon the maintenance workforce finds itself years behind current developments.

It becomes evident that the human potential is not being fully realised, at least not for today's needs. Besides being out of date the skills of workers have become stale and some abilities have become atrophied. Our first objective should therefore be to develop the potential to the fullest. What is needed is a competent "driving force" directing all activities. For it must be understood that every maintenance job, however mundane, must be performed in the light of modern technological change.

Skills are only one of the many untapped resources. Another is the human character and its aspirations. In this field of human traits and attitudes, many booster techniques have been developed in the last two decades but their tangible benefits have still to be proved. Apart from audio-visual training methods that relate mainly to supervision and foremanship few such techniques apply to workers. Programmed learning may be the only one.

Another area where much can be done is to encourage the emergence of workers' specialised talents. Opportunity should be given for people to improve their natural aptitudes and thus widen their scope.

It becomes evident that a company using all these approaches in order to make the most of its manpower potential is the exception. Few would spend the time and effort required to establish the necessary procedures and, above all, to make their supervisors and managers aware of the problems involved. It is understandable that the human angle is considered an elusive one, unmanageable and unprofitable to pursue. Ironically, monetary advantages in one form or another are offered to maintenance workers as if to compensate for their being otherwise neglected. Many authorities agree that investment in the assets of human skills is rewarding. It should therefore be given more attention by maintenance management. It has been shown that work attitudes can be conditioned by a better understanding of human needs; by a better application let us try and achieve more satisfaction for both worker and employer.

Let us now take a second look at Fig. 72, and it will be seen how vast is the number of aspects relating to this problem.

SUPERVISION

The foreman of a small maintenance group is at one and the same time a manager, a clerk, a storeman, a planner, a transporter, an instructor and a supervisor. The many duties that he is supposed to perform are essential whatever the scale of operations. The question must be asked: "Is he able to perform them?" In the majority of cases the answer must be no. Foremanship resembles motherhood in that when a girl becomes a mother, without much preparation she has to become a nurse, a teacher, a cook, an economist, a psychologist and a seamstress and, in all probability, she has had more advance training for her task than the average maintenance supervisor has had in his.

It is, therefore, little wonder that so many maintenance supervisors receive a good deal of criticism and earn very little respect, being without adequate resources and also being expected to "run a good show on a shoestring." Furthermore, after his appointment a supervisor has little time to fill the gaps in his knowledge.

The main attributes of a maintenance supervisor are:

the ability to manage people;
an adequate and up-to-date technical know-how;
the ability to plan and execute work;
the ability to initiate and sustain efficient procedures;
safety consciousness;
the ability to instruct and inspect; and
the ability to be forceful yet courteous in his relationships.

Some of these attributes can be acquired by training and experience, but this should not be taken for granted nor left to chance. The following principles must be observed:

(*a*) A time schedule for training is essential.
(*b*) A programme of studies is required.
(*c*) An understudy should be available.
(*d*) Interest must be shown by management (*e.g.* through follow-up discussions).

It frequently happens that a person returns from training with many good ideas that he wants to introduce, only to meet with objections and obstacles. As a result he becomes discouraged and frustrated. In some places only lip-service is paid to improvements suggested by staff and much more attention is given to cost-reduction campaigns instituted by management; this is sometimes called "management by crises."

It is vital to remember that a maintenance supervisor's performance determines how most of the maintenance budget is spent. Within his control lies the utilisation of time, use of materials, methods and quality of work and general conduct of the workforce. Given a team of fifteen men, a supervisor's yearly expenditure may be close to £100,000.

What qualities should an ideal maintenance supervisor possess? Rhetorically he must be able to "talk with crowds, nor lose his virtue; walk with kings, nor lose the common touch." He must be firm in discussion—authoritative without being overbearing—and he must be able to instruct and reprimand without offending.

In performing his many tasks a maintenance supervisor must make good use of his time. To ensure that everything is done, his use of daily, weekly and monthly timetables is advisable. Work allocation could be done in the mornings, inspection tours made twice a day, enquiries and correspondence attended to during the day and recording and reporting performed late in the afternoons. Meetings and conferences, though vital, can easily become a terrible burden on the schedule. It is best to reserve two alternative half-days in mid-week for these. Some supervisors have weekly meetings with their craft foremen, a practice that is to be recommended.

Weekly schedules should include the revision and follow-up of work carried over from previous weeks, preparation of weekly reports, weekly reviews of the stores and spares situation, the position on outstanding subcontracts and personnel matters.

Equally useful are monthly schedules which would deal with such matters as safety committees, suggestion schemes, budget reviews, discussions on the reordering of spares and tools and procedures within the department.

All in all the hectic work of a supervisor tends to increase sharply because the separate functions must be supervised, checked, instructed and co-ordinated. Staffing will approximate the pattern shown in Table X as the size of the maintenance groups increases.

When maintenance engineers, supervisors and foremen complain about lack of time, production or sampling studies can be carried out regarding the way their time is spent. These studies often disclose that excessive time is spent on walking, arguing, searching

Table X. Relationship of number of indirect staff to department size.

Size of maintenance dept.	*Maintenance indirect staff*
Small groups 10–15 craftsmen	Supervisor One craft foreman Half-time stores assistant Half-time clerk Total - 3
Medium groups 16–32 craftsmen	Supervisor Two or three foremen One stores attendant One clerk/dispatcher Total - 5 to 6
Larger groups 33–60 craftsmen	Plant engineer Maintenance supervisor Three or four foremen One job planner/dispatcher One stores attendant One tool stores assistant One clerk Total - 9 to 10

and repeated follow-up on work that should have been (but was not) completed on time. Closer study of the reasons for walking or arguing will probably indicate the need for better communications. Whatever time the supervisor spends on non-essential activities, reduces the time available for technical and supervisory work.

One study showed that a total of 40 per cent was spent on non-essentials. The time spent on technical instruction, supervision and discussion was as low as 25 per cent and 15 per cent of the time was used for managerial–clerical work, such as countersigning of forms, scanning reports and discussing the recording of data. Planning, out-contracting and technical correspondence took up the rest of the time, namely 20 per cent.

Investigation can also indicate whether personnel matters are taking up too much time. There are matters that should be handled by the personnel officer of the company and other more personal problems with which the immediate supervisor could deal.

Problems of a personal nature must be attended to swiftly so as to avoid grievances fermenting below the surface. However, when grievances and disputes are of frequent occurrence they indicate deeper sources of trouble and should be attended to at a different level. To obtain an over-all view, grievances should be recorded in a log book with an explanation of their origins. When their frequency rises it must be interpreted as an indication of some general malaise.

By following some of the outlined suggestions the picture of a harried supervisor frantically engaged from sunrise to sunset can be improved. Training may give him the tools to improve his practices and factory procedures can provide guidelines to assist him in his decisions. In addition, he should occasionally pause and reflect about his own attitude to his job and try to assess his performance.

INCENTIVES

The purpose and objective of incentives

In present-day materialistic life, man can best prove his worth by succeeding financially. Since the basic needs of existence are satisfied by a company's mere compliance with employment practices, a person can take his regular earnings for granted without exerting himself to any great extent. Except for an occasional challenge to one's resourcefulness, there is little encouragement for a person to apply his full abilities—unless stimulated to do so.

Motivation may come to a person either from within himself—if he tries to find adequate interest in his job—or from outside, if incentives are offered to him. Outside motivation may be due to circumstances such as premature marriage, an ailing parent or an ambitious wife. Often society will provide the incentive, such as the ambition to "keep up with the Joneses." Since in industry we have to achieve certain goals—whatever the circumstances—we cannot rely on these outside influences to stimulate workers' efforts.

If we agree with the premise that man by nature is a competitive creature, we may expect that the same spirit will also motivate within the framework of a group. To function properly, man needs a challenge. People will always try to outdo each other or to "beat the system." In other words, it is natural to endeavour to achieve success above the normal. To take pride in performing better or earning more than the next man is also a part of the same competitive spirit. An existence without this challenge appears dull and unexciting, and when no challenge exists groups will devise their own.

A situation may also arise where, in spite of the absence of an incentive, an employee has enough enthusiasm to carry him through obstacles and disappointments. However, if these in the end overwhelm him, we have a dispirited person on our hands.

In this chapter we have, therefore, to consider incentives that will enhance the competitive spirit in some people and offer encouragement to others who want to keep their enthusiasm high. Failure to provide means for such motivation will result in an indifferent workforce, performing mediocre work.[26]

We can thus sum up the purpose of incentives as offering an opportunity to exercise a natural bent for competition and to provide scope for satisfying a sense of pride and achievement. The immediate objectives therefore should be:

(*a*) stimulating the workforce to achieve better performance;
(*b*) improving group morale; and
(*c*) reducing over-all costs.

Until recently some practitioners of scientific management fostered the mistaken idea that the payment of additional money in itself will ensure a better performance. Other equally wrong conclusions about the use of incentives are still widespread today. The passage of time, and the consolidation of an industrial engineering science that can provide proper solutions, have done little to reduce the number of the many unsuitable and unscientific applications of incentives or quasi-incentives.

An example of this is the use of overtime as an incentive. The reasoning behind this practice goes something like this: "If you're a good fellow, Jack, you can work overtime. We cannot pay you extra for good performance since we can't measure it, but this much we'll do for you." In some places this practice is so entrenched that trying to remove it may cause a dispute.

By now Easter and Christmas "bonuses" have become so common that they are taken for granted. In other cases there is an effort to tie the maintenance bonus to that of production. The bonus earned by maintenance staff may be in proportion to that earned by direct production workers on the premise that if the latter are able to earn a certain bonus, it must be partly due to a better service provided by the maintenance department. This method is justified in intensively mechanised industries, but not suitable in benchwork and assembly factories.

Sometimes work relations are poor, craftsmen complain, threaten to leave, bicker and grouse, and at such times management may have the glorious inspiration, "Let's apply incentives, then they won't have time for all this nonsense!" So, another incentive scheme is introduced, superimposed on poor organisation or built on shaky or non-existent foundations. In the end it flops and some *post-factum* know-alls nod their heads and say, "I told you so."

There are many ways to motivate people. There are incentives, "disincentives" and negative incentives. The last-named are used to deter people from doing certain things. Chief among these is the fear of dismissal, which under the present almost universal protection of trade unions is very remote. Since psychology has been widely adopted in the conduct of industrial relations, public reprimand has been replaced by a discreet reproach in private. As fines and penalties are outlawed, there is, therefore, little left in the arsenal of negative incentives. An effective method of motivation, however, still remains in this category, namely social ostracism. When within a certain group an individual is evidently not doing his full share his colleagues will often exercise a certain pressure on him in most subtle ways or exclude him from their midst. It is up to the company to foster a team spirit strong enough to adopt this approach. But here again we cannot rely on this to happen.

Some incentives aim for short-term and some for long-term effects. Some may attract workers to the workplace, but these may not motivate them to exert more effort *during* work. Thus we have to review the whole range of incentives in order to be able to choose the most suitable ones.

How can we motivate?

"Disincentives" or work deterrents are plentiful. Their effect is to reduce the will to work and distract employees from their immediate tasks. A few are the fault of management neglecting to provide certain conditions that workers consider essential in their daily duties. Workers are thus discouraged to make extra efforts in order to overcome what they consider to be deficiencies. Poor tools and shop facilities, a slow stores attendant, insanitary working conditions, etc., can act as very effective work deterrents. Other factors can also impede work. On the day of a great sports event, for instance, all you need is a remark about those lucky fellows who can go and watch the game and output will be minimal. Such events are strong deterrents but outside our control.

In some cases incentives are applied in order to outweigh as far as possible the inhibiting factors—the irritants. As long as we do not remove them we have more obstacles to overcome; the more prevalent they are, the more we need positive incentives and the more costly it becomes to the company.

Now let us assume that in the management sense we are right and we choose the right incentive. The intangible factors of the workforce, its background, its inherent motivation, the informal structure of the team, their age and aspirations, begin to play a part. Regardless of how sound our proposed system may be, it cannot take everything into account. The problem of incentives therefore means a *removal of all deterrents to work* and the *introduction of a work-appeal* to satisfy the *majority*.

Adopting the wrong incentives can be compared with playing an organ to charm an Indian snake. This is the problem of incentives in a nutshell. To reduce the chances for costly misapplications, an effort should be made to find out what aspects seem most important to the maintenance crew, and having done so, set up the right incentive measures.[37]

Motivational factors

Motivational factors have been identified by recent investigations and classified into the following basic groups:

(*a*) *Subsistence and safety needs.* The needs for food, clothing and housing which set the level of material wealth, and the need for security.

(*b*) *Social needs.* The factors that relate to a person's standing in the society that surrounds him.

(*c*) *Self-esteem and fulfilment.* The need that gauges an individual's attainments in respect of his moral code and aspirations.

In trying to present these incentive factors to a group we soon discover that the needs vary with the individual, his education, the circumstances of his family life, and that they are strongly related to the prevailing economic conditions. All these are a function of time, namely they rise or pass as time goes by, as a person becomes older or a country's fortunes fluctuate. Even when social and economic conditions are considered relatively stable—a highly unlikely assumption these days—maturity and age will cause shifts in a person's needs. Thus any set of factors will affect a group of people in different ways, and poses serious practical problems.

1. In broad terms, we assume that everybody has the same ECONOMIC ambitions, the affluent society being the outcome of the urge to acquire and to possess. This aspect, however, is not universal. Since the advent of world-wide mobility and of closer international ties, it has been discovered that in certain societies the acquisitive spirit is not dominant. Even in economically advanced countries the materialistic urge is far from uniform—the values differ. Therefore, although money *is* a motivational factor it does not apply equally to everybody.

2. SOCIAL values within a certain society having a homogeneous cultural background appear to be more stable. Most men, barring extreme personalities who in one way or another shun social contact, will prize their position in society. Whether we like it or not, people are made aware of their status by those around them. A position, however modest, carries certain exclusivity that gives it status. An office messenger who carries the letters for everybody has his special status. Similarly, the night-watchman who is in supreme control at night or the boiler attendant who is king in his domain.

Status recognition comes not only by appointment. Professional competence earns a reputation, and personal services to a group earn recognition. These qualities lend a standing in an informal way. The social framework of any group is a complex pattern of relationships which usually places individuals according to their merits. The better the positions in a social framework are defined, the greater is the importance of these factors to the individual. Conversely, a group with loosely defined relationships will breed apathy, discontent, friction and ultimately revolt.

These conclusions support and justify the principles of good management, namely, the establishment of an organisation chart and the need for job descriptions. When these are lacking the motivating value of social factors diminishes.

The often misused term, "human relations," acquires a different meaning when it is realised that relations are mostly the outcome of simple procedures, such as job descriptions, since they are a reflection of status positions. Without attempting to deprecate the importance of courtesy and human understanding which is mainly implied by the popular term, "human relations," the following simple sequence seems perhaps more important:

RESPONSIBILITY requires AUTHORITY;
AUTHORITY establishes RELATIONSHIPS;
RELATIONSHIPS determine STATUS; and
STATUS sets mutual ATTITUDES and reciprocity.

We thus have scope for providing the social motivational factors without incurring excessive costs, except to make the effort required in establishing satisfactory procedures. There is nothing highfalutin about this.

3. Factors of SELF-ESTEEM and FULFILMENT are very subjective, and if we agree that they result from education and social climate, these factors can be enhanced. In this respect, employment plays an educational role, since it offers people the opportunity to socially achieve self-esteem. The better that role is carried out, the more the company can expect in return. Again, these factors do not appeal equally to everybody. Much depends on individual personalities.

Thus, employment that tends to satisfy the economic, social and ego needs of the employee, should be considered rewarding enough for management to expect maximum efficiency.

All measures that lend "work appeal" (not mere "clock-in appeal") to a place can be used as incentives. Surveys have indicated that, notwithstanding the earlier grouping of a person's needs, the factors shown in Fig. 75 are of foremost importance to employees.

(*a*) *Security*. The prospect of continued employment.
(*b*) *Acceptance*. The awareness of favourable relationships.
(*c*) *Recognition*. The feeling of being appreciated for one's contribution.
(*d*) *Interest*. The degree of involvement in work and the resulting stimulation.
(*e*) *Progress*. The chance to learn, promotion and the expectancy of an improved standing.
(*f*) *Remuneration*. Absolute *and* relative earnings.

The results found by surveys covering these aspects vary considerably between periods of time and locality, but they dispose of the notion that people work mainly for money. In fact, the earliest experiments in industrial psychology, the Hawthorne experiments in 1927, disproved that assumption. Ever since, this branch of studies called "human relations" has attempted to clarify the issue.

Types of incentives

Technically speaking then, there are non-financial, semi-financial and financial incentives. Figure 75 shows a broad range of incentive measures. Non-financial factors must be *inherent* in the job, while semi-financial factors are *offered* as opportunities to derive satisfaction from or to complement job aspects. Financial schemes, on the other hand, are *procedures* that must be set up and followed. If we allow Fig. 75 to serve as a checklist, we may select those factors that seem applicable to our own circumstances.

It would be unwise to ignore offhand the non-financial incentives although they are more difficult to apply in practice and their resulting benefits are more complicated to discover. Nevertheless, in the attitude of maintenance workers we can identify a number of factors which provide satisfaction, attraction and compensation to the craftsman. The sense of pride when he completes a good job, the trust that is placed in him, the company's dependence on his work, etc., are not negligible factors, however intangible they appear. Nevertheless, he can hardly spread these on his bread. Company prestige is, however, undisputedly a great source of pride, although not everyone can work for a world-renowned firm.

Semi-financial incentives are not less important. Company training programmes, suggestion schemes, improved canteen facilities and better working conditions, all contribute to a more satisfying background for work. In the end, however, these cannot

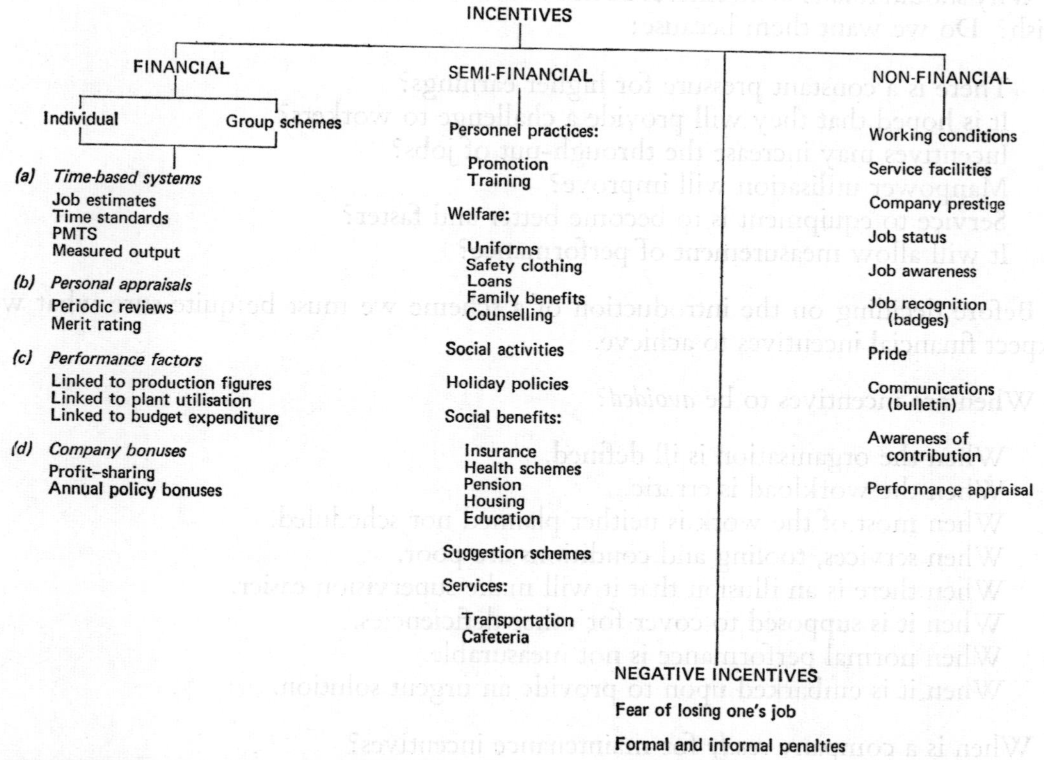

Fig. 75. Classification of incentives—financial, non-financial and negative.

be considered as inducements to work harder while on the job. The reasons why these practices may slightly boost a person's performance is *not the effort made to obtain the benefits, but rather the fear of losing them.* He will sometimes perform so as just to keep himself out of trouble or from being dismissed. This "brinkmanship" is not really what is wanted. Let us now consider financial incentives where most of the attention is usually concentrated.

Financial incentives

We define a financial incentive as "extra payment earned for performance above normal." This is not a perfect definition but it can be understood. The trouble lies in the word "normal" because that level may mean anything between a 40 and 80 per cent time utilisation. We cannot call it "average" because that value fluctuates. In a good group the average may mean 30 per cent more effort as compared to another team. "Extra" is also ill defined and often disputed, and "payment" can be claimed in cash, in kind or other benefits. When considering the introduction of an incentive scheme these terms have to be clearly defined. Before we go any further some questions must be answered in full:

Why should financial incentives be introduced? What do we expect them to accomplish? Do we want them because:

There is a constant pressure for higher earnings?
It is hoped that they will provide a challenge to workers?
Incentives may increase the through-put of jobs?
Manpower utilisation will improve?
Service to equipment is to become better and faster?
It will allow measurement of performance?

Before deciding on the introduction of a scheme we must be quite sure what we expect financial incentives to achieve.

When are incentives to be *avoided*?

When the organisation is ill defined.
When the workload is erratic.
When most of the work is neither planned nor scheduled.
When services, tooling and conditions are poor.
When there is an illusion that it will make supervision easier.
When it is supposed to cover for other deficiencies.
When normal performance is not measurable.
When it is embarked upon to provide an urgent solution.

When is a company *ready* for maintenance incentives?

When control procedures are satisfactory.
When industrial relations are not explosive.
When human relations are harmonious.
When communication channels are working.

How is the introduction of incentives to be approached? Principally, it should be *with care*; in addition, by talking it over repeatedly with all concerned.

The pros and cons of the situation must be carefully weighed. The cost of introduction is a major factor, both in the matter of expense and the time that it can be expected to take. The implications of incentives go further than we often realise and this must also be considered. The stores and boiler attendants may also, rightfully, demand that they be included in the scheme. Then, how about maintenance foremen and supervisors? Above all it must be clear that whatever the system, its introduction is not always a smooth process.

The introduction of a financial incentive must therefore be viewed with all its implications. The scope of the scheme must be foreseen so as not to create inequities among crews and individuals—this is the sort of issue that can create crises. If foremen and supervisors are not included it may in due course cause resentment and dissatisfaction. On the one hand, care must be taken not to create tensions; on the other, there must be a return on the money spent on incentives.

Incentives must also be directed in the proper way towards the main objectives. If too much emphasis is placed on time-saving, shoddy work may result. If there is excessive encouragement to save on parts and materials, the result will soon be apparent. The best insurance against committing costly errors is to be well informed about all the aspects and all possible alternative solutions. Another way to insure against unpleasant surprises is to employ an experienced consultant.

The period of implementation is usually longer than anticipated. Difficulties in measurement may arise and the coverage of work with job times, or the establishment of 100 per cent levels or bench-marks, may at times proceed very slowly. At a later stage recruitment of the proper administrative staff may pose problems. Still later, the introduction of control procedures may give rise to difficulties.

This wide range of problems is different with each incentive application. An over-all consideration that must not be overlooked is that a basically *good wage scheme* is an incentive in itself. Conversely, *a financial incentive should at no time be used to correct an inadequate wage structure,* nor to overcome wage inequities.

Now that the difficulties of implementing an incentive scheme have been presented, let us turn to the brighter side. Fortunately a wide range of alternative schemes is available and companies can choose one or a combination of schemes that suits them best. For reasons of size, organisational structures and operational characteristics in different departments a variety of schemes may be in operation in the company. Thus we may have group schemes or individual measurements, time-based schemes or a system tied to production performance.

Bases for financial incentive schemes

Assuming, then, that the introduction of an incentive scheme is found to be justified *and* feasible, then the pros and cons of the available alternatives may be examined:

(*a*) *Time-based systems*
Job estimates.
Time standards.
Predetermined motion time standards (PMTS).
Measured output.

(*b*) *Personal appraisals*
Quarterly reviews.
Merit rating.

(c) *Performance factors*
Linked to production figures.
Linked to plant utilisation and downtime.
Linked to budgets and expenditure.
Linked to total maintenance costs.

(d) *Company bonuses*
Profit sharing.
Annual policy bonuses, *e.g.* 13th month.

Time-based systems assume that the time taken to perform a job is a good measure to assess the level of individual effort applied and thus indicate the eligibility for a financial award. Since craftsmen are being paid for the time spent at work, and their services are always in demand, it seems natural for management to make the most of their time. This is the reason for emphasising time performance.

Although this approach seems logical, practical *and* feasible in some companies, it is not completely watertight. A worker can produce excellent results as measured by a time-scale and still his output may only reflect half of his *total* abilities, or possibly the job may not have been performed as thoroughly as it should have been. A time-based system will achieve most when it is understood that the company regards this incentive as *a measure for the possession of skills and know-how and a show of devotion* to the employers: rushed and superficial jobs are not to be condoned. To get these points across is not always easy, and in most instances communications are not good enough to achieve this. Thus placing the emphasis on time-saving may become a mixed blessing.

In addition to this drawback, the procedural demands of time-based systems are quite exacting: every job has to be estimated, either before or after completion; up-to-date reference tables for all operations must be developed and maintained.

Schemes have been devised to reduce the requirements of accurate estimates and to support teamwork while rewarding personal endeavour.

The procedures involved in time-based systems are not complicated, but the following points are fundamental:

(a) Time standards are necessary for most jobs or their elements.
(b) A time-setting procedure must exist.
(c) Job allocation and despatching must operate smoothly.
(d) Recording and controls are necessary.

When the emphasis falls on time performance, the lack of available services may be highlighted. Since a time standard assumes certain conditions, pressures are created *for* certain things and *against* certain practices. The pressures for showing good performance may also give rise to friction in other areas, which are not really in a condition to support work done at incentive pace, such as collaboration between crafts and proper service from the stores. Other effects may be related to supervision and teamwork. Where supervisors are lacking in ability, and teamwork has never been a strong point, a time-based system will emphasise their failings.

When measures are taken to ease these pressures time standards may need to be revised. Take, for instance, stores service—whether tools, supplies or spares. If at the start service was poor, and this was allowed for in the time standard, a subsequent improvement may cost money to be implemented and also cause the previous standards to become "slack" or obsolete. The costs of administration of all time-based systems are higher than for others, and a great deal of their success depends on competent estimators who are always in short supply.

On the other hand, schemes that are based on individual appraisals such as merit rating are relatively easy to administer once agreement has been reached regarding the choice of factors and their relative importance.

The bonus is calculated according to the decision of a small panel of assessors. The panel usually consists of the immediate supervisor, a representative of the workers and a member representing technical operations.

The panel must meet at regular intervals to assess the performance of each worker on a suitable sheet. This form lists the factors that are to be rated and includes the rating achieved during sampling studies; the total figures arrived at are the basis of the calculation of a bonus. Assessments of a series of personal factors should be made either bi-monthly or quarterly and payment should be prompt.

This system is simple and effective, although the tendency is for assessments to rise and rarely to fall—after a while everybody earns maximum bonus. Crew size is another limitation. Beyond a certain limit the job of assessment becomes tedious, even if performed only at quarterly intervals. It can be argued that the incentive power decreases with increasing length of time between payment periods; however, the accumulated amount may compensate for this disadvantage.

Next to be considered are incentives based on performance factors of various departments of the plant. Factors must be chosen that are affected by the work of the maintenance crew and there must also be adequate and reliable procedures to obtain monthly results. These schemes, being group schemes, reduce the individual impact of an incentive scheme. However, they emphasise a collective responsibility of the group and are more attractive to management, and are not costly to run and install. To obtain the maximum benefit periodical discussion meetings are advisable, where variations in results can be queried. Two examples of such schemes are given in Fig. 76.

Cost reduction *per se* should not be used as an objective, since costs could be reduced below the figure which can sustain a reasonable level of service. These schemes also require a background of smooth work-relations and must be supported by sampling studies and supervisors who are themselves capable of high incentive performances.

Company bonuses, based on profit-sharing or annual dividends, are rather distant when viewed from the repair workbench. They are not specific solutions for our needs in maintenance, and rather serve to improve the company's image to its workers.

In summarising this topic, we can say that while incentives *are* applicable and useful for the maintenance department, they are neither absolutely essential (if a proper wage scheme is in existence), nor are they easy to apply and maintain. They *do* produce results—at a certain cost. They are certainly not to be used as palliatives, stopgaps or emergency measures.

	Job group: Description	Average time	Jobs completed this week	Total hours earned	Total direct hours recorded	Efficiency level, R	Rating factors (%) for different efficiency levels: R70%	R75%	R80%	R85%	R105%	R110%	Credit hours	Bonus calculation
	a	b	c	d=b×c	e	f = d/e							g=dxR	
1.	Minor jobs 20 min–1 hour	0·67 hours	20	13·4	Accumulated from job-cards and/or foremen's daily reports	482·4 / 562·3	101	102	103	104	108	109	13·7	
2.	Small jobs 1.1–5 hours	3 hours	18	54			108	109	110	111	115	116	59·0	604·1 − 482·4
3.	One-day jobs 5.1–11 hours	8 hours	10	80			115	116	117	118	122	123	92·6	= 121·7
4.	Two-day jobs 11.1–21 hours	16 hours	5	90			122	123	124	125	129	130	110·8	at avg. hourly rate
5.	One-week jobs 21.1–50 hours	35 hours	3	105			129	130	131	132	136	137	136·4	× £0.85
6.	Project class 50·1 and above	70 hours	2	140			136	137	137	138	142	143	191·6	= £103.50
				482·4	562·3	R=76%	Range and scales of rating factors according to recent averages and accepted target figures						604·1	for distribution

Procedure for calculating INCENTIVE BONUS BASED ON WORK COMPLETED

	The factors: Description	Relative weight	Maximum % of bonus attainable	Range of monthly results and corresponding rating factors	Example: Nov. '71 score: Perfm.	Rating	Credit	Bonus calculation
		I	II		III	IV	(IV x II) %	
(a)	Plant utilisation	40%	14	Perf. 85% 86 87 88 89 90 91 92 93 94% rating % 70 75 80 85 90 95 100 105 110%	91	95	13·2	30·2%
(b)	Output related to target	30%	10	Perf. 80% 85 90 95 100 105 110% rating % 80 84 88 92 96 100%	100	92	9·2	at avg. hourly rate × £0.85
(c)	Scrap and waste	20%	6	Perf. 8% 7% 6% 5% 4% rating 64 76 88% 100%	7	64	5·8	bonus per hour
(d)	Indirect hours	10%	3	Perf. 5% 13 11 9 7% rating 40 60 80 100%	7	100	3·0	= £0.255 on attendance
		100%	33	Performance ranges to be established from currently attainable figures	This month's score		30·2	

Procedures for calculating INCENTIVES BASED ON PLANT PERFORMANCE FACTORS

Fig. 76. Two procedures for calculating group incentives based on work completed or plant performance.

THE MAINTENANCE MANAGER'S JOB

Who is the right kind of person for the job of maintenance manager? Where do we find him and how do we retain his services? Alternatively, can we make the present occupant of the post more efficient? To answer these questions, the nature of this job must be clearly understood.

Little attention has been devoted to assess the requirements of a good maintenance manager. For years, industry has wrongly assumed:

(*a*) that a good "technician" can also become a good team-leader in his professional activities; and

(*b*) that managerial abilities can be developed by just practice and experience.

It was often taken for granted that given adequate seniority, and with a certain amount of training, foremen can become supervisors and supervisors can rise to become good managers. A reasonably good manager in one field was expected to adapt himself to the challenge of other positions to which he was transferred. In other words, whether an experienced supervisor was promoted or whether the manager of another department was given this task, we could expect equally good results. In the light of more recently defined knowledge from the behavioural sciences, these assumptions have been proven wrong. Management is a complex activity which has often been described as an art because of the many elusive and undefinable elements in it. Personal qualities, knowledge, skills and experience all merge into the practice of management. Moreover, the practice of management is as particular to individuals as it is specific to certain positions in different companies.

There have been many books on the science of management, and in the process the subject has become vastly ramified. Management has been defined, among other things, as a composite of forecasting, planning, organising, directing, co-ordinating and controlling. Other definitions combine the terms resources, objectives, goals, policies, organisation and others. Although these definitions reflect the technical–professional aspect, they do not even pretend to touch upon the human implications of management.

It is evident that since the manager is human and since he accomplishes his task *with* the assistance and *through* human efforts, his activities and the results he is getting are greatly affected by human aspects: his purely technical work can never be divorced from its content of human strength and weaknesses. Efforts have been made to prescribe what a manager is expected to do in expressions such as "anticipate trouble," "keep a broad perspective," "work towards an objective," "inspire confidence," "set a good example," etc. These expressions endeavour to combine the technical with the human aspect.

Therefore the best course of action for our purposes appears to be in defining the technical *and* the human aspects of a manager's job. Table XI lists the main ingredients.

If we agree that a certain "mix" of these ingredients can make up the widest range of managerial functions, let us see in what particular ways a maintenance manager's job differs from others. Obviously, filling this post with the right person is very difficult, and unless he is the right man his life will be full of frustrations.

Volumes have been written on leadership alone. Because of frequent semantic problems it appears that the term "good leader" has become synonymous with "good manager" and vice versa. This is not necessarily true. Many good leaders are very bad managers. It is undisputed that a good leader must have a flair for organisation if he is to retain his position for any length of time. He must also be able to make decisions—the majority of them the right ones—to prove his ability. Within the maintenance

Table XI. Leadership and managerial abilities required in a maintenance manager.

THE MAINTENANCE MANAGER	
Characteristics of a good leader +	Traits and skills of a good manager
Personal qualities:	Organising ability:
Honesty and courage	Ability to delegate
Concentration and persistence	Methodical approach
Confidence and reliability	Effective use of time
Initiative and resourcefulness	Planning ahead
	Effective implementation
Human relations:	Analytical ability:
Courtesy and firmness	Discerning priorities
Helpful and instructive	Problem-solving
Ability to listen	Decision-making
Leadership 'style':	Management know-how:
Team leadership	
Motivating people	Up-to-date techniques in technology and management
Judgment and fairness	Communications
Power of expression	Utilisation of resources
Sense of responsibility	
These characteristics determine how a manager deals with people	These traits and skills reflect his ability to solve problems

organisation, leadership without technical know-how is unthinkable. Since organisational ability, decision-making and other abilities are components of management and not necessarily inherent in leadership, we can also conclude that some good managers may not be very good in leadership.

How does this discussion apply to maintenance? In simple words, to satisfy the company's best interests we need a maintenance manager who is both a good manager and a good leader. As described in the early part of this book, the maintenance function incorporates many activities of a "manufacturing" enterprise. It is a conglomerate of technology, people, economics and management (forecast, plan, organise, direct, co-ordinate and control) unlike any other. It is therefore very demanding in terms of managerial ability. The ability to steer a middle course between *workers and management*, between *the service needs of plant and the economics of the enterprise* as well as between *professionalism and the realities of a situation*, these are the pressures that create a terrific strain on maintenance managers.

The ideas expressed here reflect the recently developed "Management Grid" and the "3-D Theory of Managerial Effectiveness".[42] Both theories claim that there is an infinite variety of personalities and the way they perform falls somewhere within a certain range of technical and human qualities. Both systems assign numerical scores and descriptive terms to the analysed individuals. Although helpful and enlightening they are difficult to apply in practical terms.

A discussion on all the points listed in Table XI would be too lengthy for this volume and only the more salient characteristics are here discussed.

Leadership refers to the manager's way of handling his staff: human traits and qualities as well as his managerial "style" determines a person's standing within his group. All problems, whether technical, financial, organisational, etc., must be resolved through communications and human contact. A sound technical judgment does not obviate the necessity for the right human approach. Inherent in leadership is a sensitivity to people, their needs and problems, and an understanding of how to motivate them and bring out the best in them. Leadership means a good degree of "comradeship" (friendly and genuine interest), but without an excessive familiarity that would destroy the status of formal authority.

In some books a distinction is made between leadership qualities that are acquired and those that are innate. The latter include the exercise of judgment, analytical sense, attitude towards people and adaptability: while these can be developed by training they cannot be implanted from the outside. Traits that are acquired and can be developed to a certain extent include the ability to communicate, to listen and to handle people's problems. These are necessary components of a good leader and much can be done to encourage people to develop in the desired direction. This can be achieved by giving them enough authority to exercise their leadership abilities and learn from experience. Alternatively, participating in manpower development programmes is advisable.

The subject of leadership is further complicated by distinguishing between *know-how*, *conduct* and *attitude*. Supervisory training and foremanship use this approach to develop leadership and managerial abilities. "Training" can be accomplished by a variety of methods ranging from lectures and group discussion of case studies to psychological methods such as sensitivity training. While all these are useful much can be said in favour of good old-fashioned reading: some reading programmes are available from specialised training institutions which provide printed articles, checklists and other useful material on a subscription basis. Much can also be said in favour of membership in professional societies where an informal exchange of experience takes place.

Organising ability is an important quality of a manager. To organise his own activities and that of his staff is one visible way of proving his managerial abilities. This, of course, is more easily said than done. A person will often feel, especially in a maintenance department, that he is at the mercy of many circumstances beyond his control.

After a number of years in this position, unless he is a person of outstanding resilience and of utmost determination, he may be reduced to shrugging his shoulders, and being resigned to his fate. The possession of good organising ability may improve matters, but only stubborn persistence can ensure lasting results.

In practical and tangible terms, you must be able to organise your desk, your paper-work, your time and your contacts. You must be able to organise the tasks of your staff effectively, so that their duties are both permanent and flexible.[39] You must be able to assign priorities, lay out a sequence both for long-term projects and for day-to-day operation. Lastly, you must be able to deal with emergencies without faltering. Quite apart from the necessity to employ principles and techniques of organisation as defined in textbooks, it appears that good organisation stems from another quality of leadership, namely *discipline*.

Decision-making abilities can be developed by practice and by mastering certain helpful techniques. A common problem here is the fact that having carried responsibility for years many supervisors and managers have acquired a certain set way for making decisions which may be quite difficult to change. Then, of course, the expression is heard "you can't teach old dogs new tricks," which expresses the resistance that will be encountered against change. Nevertheless, group training is particularly helpful in this area.

Decision-making is greatly helped by explicitly stated company objectives. This reinforces remarks made earlier in this book that a systematic approach starting from policies and objectives is necessary in order to allow staff to operate satisfactorily. Procedures of planning and control transform objectives into management data on which decision-making is based. Thus, we again see how one factor supports another, and one link in the chain leads to another. It is evident that a good decision-making ability would be lost without possessing the prerequisites of objectives and management data.

Systematised decision-making techniques have been developed involving mathematical models and numerical evaluation of factors. Among these are linear programming, simulation and aspects of the probability theory. Except in the case of medium- to large-size groups a maintenance manager would not have much opportunity to use them. And in cases when these could be used, an analyst will usually be employed for the purpose. A familiarity with the techniques will nevertheless be advantageous.

A maintenance manager's need for technical know-how refers as much to technology as it does to management. Although he need not be an expert craftsman he must be sufficiently conversant with the technology of the plant to be able to support the work of his staff by finding and providing assistance in tooling and information whenever help is required.

To a great extent it is the company's responsibility to provide the proper opportunities to its managerial staff to improve their performance and prepare themselves for executive positions. Similarly, executives themselves should realise that further development of their abilities is in their own interest and that of the company. Being well informed on the available means for training is therefore necessary both for top management and for the executives themselves.

The task entrusted to the maintenance manager makes it imperative that we maximise his performance—even a slight improvement in his work might save the company a considerable amount of money. To achieve this he must be allowed time for participating in development programmes, whether within the company or outside. A good way to do this is to provide an assistant-cum-understudy who would fill in when the manager is away. The extra expenditure will prove to be well worth while.

Chapter 10

Control procedures

The necessity for controls in maintenance arises with the growth of an enterprise and its increasing complexity. It is a part of the stream of control information that flows throughout the firm. The main problem of the owner–manager in companies where he is the central authority is that of keeping himself well informed and following up on issues that need his attention. With the increasing size of his company there comes a moment when he must delegate some of his authority by adopting policies and giving directives, whether written or verbal. Facts that were adequately clear for him to act upon must now be explicitly documented to enable action to be taken by others. Controls that the owner–manager could carry in his head or in a small notebook, now require recording procedures that will tell him what has been done, so to speak, in his name. When a team of managers have to collaborate they also need figures to guide their decisions and to accept subsequent responsibility.

This development means that a number of basic prerequisites are essential if the controls are to be effective:

(*a*) The right data must be collected using prescribed procedures so that they will remain consistent between periods, and in order to reflect similar activities in a uniform way.

(*b*) The method of compiling data must be uniform so as to allow of a systematised presentation.

(*c*) Information collected must be relevant to the objectives and policies of the company.

(*d*) The data must reflect the variability of factors which are to be controlled.

(*e*) The personal factor should not be allowed to affect the collection, tabulation and interpretation of data. In other words the significance of data must be self-evident, provided they are properly presented.

The emphasis must be placed on the need for homogeneous, impersonal and relevant data. By comparing results with pre-set targets the manager should be able to draw reliable conclusions upon which he can base his actions.

The maintenance manager on his assumption of authority, requires controls to enable him to run his department efficiently. What follows here is a framework providing the maintenance manager with the means to exercise control over his department's activities. First it must be said that the best controls are not necessarily the most elaborate ones; often the simpler a control procedure is the more useful it proves to be. Furthermore, not all the measures described here have to be applied concurrently in order to make a system complete. The exercise of controls has to conform to the task in hand. A wide range of measures is available in order to achieve the best effect. An elaborate procedure will usually better serve a larger and more complex company where the sums at stake are also greater. Thus a small company can achieve satisfactory controls in a simple way as long as it preserves all the ingredients of a complete control cycle.

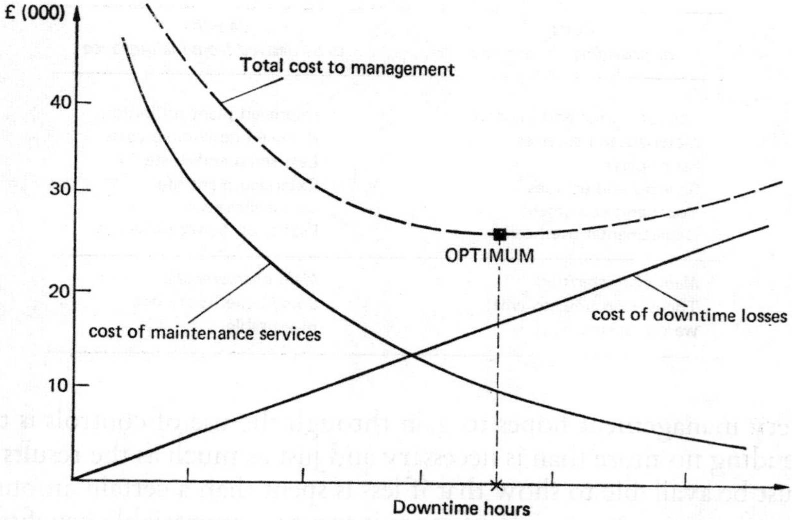

Fig. 77. At the optimum point for total costs to management a balance between the cost of losses and the cost of providing the services is obtained.

The term "control" implies:

(*a*) Receipt of adequate information reflecting the conduct of operations.
(*b*) Ability to compare results with pre-set targets.
(*c*) Mastery over situations and ability to take appropriate corrective action.

These are the conditions that any control worthy of the name must meet. How does this apply to maintenance? We must control those aspects of maintenance that are most important in achieving our objectives.

Maintenance service is provided to satisfy the needs of the plant, therefore we first concentrate on "*controlling*" *the effects* of the service on the equipment serviced. The ability of the maintenance department's chief executive to manage his activities is reflected in the way *the service is organised and performed*, and in the cost of providing the service. This is the second objective. The over-all *balance between the benefits and the cost of the services* is the third aspect that should be controlled. These three variables have been brought together in the now well-known curves shown in Fig. 77.

Apart from its value in reflecting historical records *and* as an empirical tool to determine the optimum point of operation, the chart shows that there must be a balance between the expected benefits and the cost at which the service is operated. When we try to economise in expenditure the cost of breakdowns will rise and if we invest too heavily in providing the service, the returns on expenditure may not be justified by the costs. In other words, we can achieve economy by pursuing both ends simultaneously. A list of costs and benefits is shown in Table XII.

Table XII. Costs and benefits.

Costs of providing the services	*Benefits to be derived from maintenance*
Labour, direct and indirect	Improved plant utilisation
Materials and supplies	Reduced down-time costs
Spare parts	Less scrap and waste
Services and utilities	Extended plant life
Tools and equipment	Less accidents
Departmental overheads	Profits on better deliveries
Main characteristic: There is no limit to what we can spend.	*Main characteristic:* Some benefits are not measurable.

What every management hopes to gain through the use of controls is the certainty that it is spending no more than is necessary and just as much as the results warrant. A measure must be available to show that if less is spent than a certain amount, the losses will rapidly increase, or conversely, if more is spent no appreciable benefits will accrue. At the same time the expenditure on maintenance must be well spent. This is the reason for exercising controls on the three main aspects defined above.

Control of maintenance is expressed both in terms of money and in operational figures. The latter reflect the performance of the maintenance team and that of the plant being serviced. Although money is considered a common denominator it cannot quantify variables of operation and therefore other units of measurement must also be used.

To provide the required controls, data must be gathered, analysed and presented in a concise and meaningful manner. Analysing the data is an essential part of controls and perhaps the most often neglected one. It requires the inspection of figures and the identifying of deviations from the norm, in terms of magnitude, frequency and grouping. Interpretation is a further step that endeavours to find the reasons behind the results and discovering an indication to possible remedies. Identifying trends in figures by plotting them on charts will be found to assist interpretation.

THE FLOW OF CONTROL DATA

What are the required data? What are the essential figures that we need? Are they readily available? Where do we get them from?

First, it must be understood that control data is expressed in values and figures (*see* Table XIII). In some systems verbal assessments may be used, where the terms *excellent*, *good*, *fair* and *poor* are equivalent to numerical values. Even so, this approach may be open to criticism for being subjective, since a verbal rating reflects the personal approach of the individual doing the rating. It is often biased by the scales chosen, which may be arbitrary. Therefore this approach should be avoided.

Table XIII. Measurable units of control data.

Relating to	*Costs*	*Hours*	*Frequencies*	*Other units*
Maintenance	Total costs: materials labour spares tools services overhead costs	Total hours Overtime Scheduled hours Craft hours Direct/indirect	Breakdown analysis	Orders processed Issue of spares
Production	Breakdown losses Rework costs	Direct hours of production workers and machines	Number of breakdowns	Units of output Quality and scrap %
Plant	Total investment Overhead costs	Utilisation, hours lost	Completion of overhauls	kW–hours HP connected

Secondly, control figures are generated naturally by the activities in progress. The question is whether the right details are being recorded. Is the recording accurate? Is it regular and consistent? Is the data being transmitted? If these conditions are met, we can proceed.

Thirdly, the data must be assembled so as to be meaningful.* In its progress from shop floor to management it must serve different levels and must accordingly satisfy those parameters that require interpretation at each level. The reader is here referred back to Fig. 27.

Fourthly, the absolute values cannot be used to interpret a situation. The fact that

* One is reminded of an interesting example used in TWI Supervisory Training: On a sheet of paper figures ranging from 1 to 99 are typed in what appears at first to be a completely haphazard scramble. When the sheet is folded in half horizontally, a pattern emerges. All the figures above the fold are odd, and those below it are even. Then the sheet is again folded in half, this time vertically. Now the trainees discover that each quarter of the sheet contains a specific range of figures, namely from 1 to 25, 26 to 50, 51 to 75 and 76 to 99. This proves that figures appear to be in a jumble until we apply a system to interpret them.

£500 was spent in one month on maintenance in a certain department is inconclusive by itself. This figure becomes meaningful when related to the past or when it is expressed as a ratio of other significant figures, such as an investment of £100,000 in that department, or to the fact that fifty workers worked there a total of 9,000 man-hours during that month which was 10 per cent more than last month, and that output from the department was £50,000. It may also be relevant that in a comparable month last year 12 per cent less was spent there on maintenance. One may also want to compare these figures with those relating to other departments.

Lastly, the exercise of controls is not a one-way street and should not be the result of a rare spurt of decisive management action. To be at all indicative the exercise of controls must be a continuous activity. Since it may indicate weaknesses in a host of different functions, we must be prepared to draw conclusions and to act upon them, even though it may require "stepping on somebody's toes." However, if no corrective

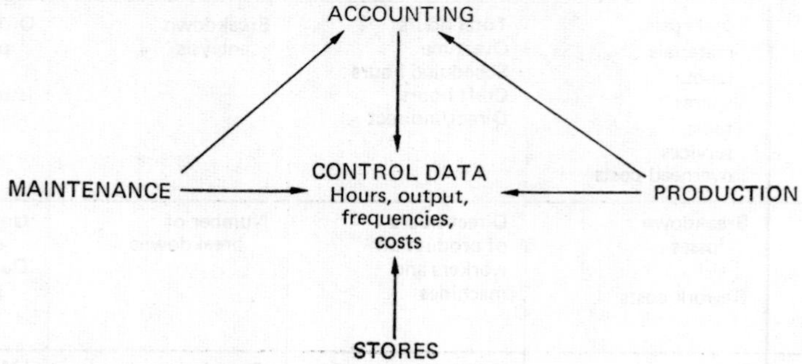

Fig. 78. Sources of control data and their interrelationship.

action is taken for whatever reason, the term "control" soon becomes a joke and nobody will take it seriously. When that happens poor results become the accepted norm and people relax their efforts. In other words, let us not try to discover difficulties or deficiencies if we do not intend to follow them up by corrective action.

In Chapter 5 we described shop controls that assemble the data at first hand, on the shop floor. The basic facts as they exist there are the only pure facts. As soon as average costs per hour or other computations are made for different crafts or machines, the original data becomes distorted. There is no overhead allocation system that is fair to all functions at all times, however conscientiously compiled. Overheads blur our ability to assess values. For instance, if it is a fact that a ball-bearing cost £2·00 when it was purchased two years ago, charging £3·10 for it—because of storage costs and depreciation of *other* items, the handling cost of *other* items, or rise in the cost of living factor—is pure conjecture or just an accounting device. Because of this transformation of data, it is advisable for maintenance to deal with information as it is obtained at first hand. The sources in Fig. 78 provide relevant data although the form in which they are used should be strictly defined.

As already mentioned, figures have to be examined not as to their absolute magnitude but in light of other figures. There is a mutual interdependence which is important, in fact highlighting these relationships is one of the main tasks of controls. This is the reason for requesting figures from Accounting and Production. The activities of Maintenance must be viewed in light of the other data. Let us consider in detail what data is available at each of these sources.

Production management collects information on output at various points—quality and scrap, utilisation of manpower and machine downtime—in other words, the "volume of work" done. The results achieved in production departments are affected among other factors by the maintenance service and we wish to discover whether maintenance has contributed to a better production performance. We will endeavour to identify this contribution.

Accounting, and accounting only, can furnish us with summaries of expenditure incurred to meet the needs of maintenance such as consumption of tools, power, spares, labour and materials. The accumulated cost charged to cost centres is also available. Meaningful control is only possible when the itemised costs are defined and related to their correctly identified account.

This, unfortunately, is not always the case since accounting figures sometimes seem to serve different needs. A common failing of expenditure controls is that the accounting function classifies costs according to their own ideas, which do not always coincide with those of maintenance. Unless detailed procedures and an account numbering system has been agreed upon *and* is being adhered to, the budget and the summarised expenditure items never agree. These are some of the reasons for discrepancies. What is to be included in each account must also be laid down in a procedure.

To illustrate this point we can take a certain department containing several groups of machinery. It may be that according to accounting procedures, compressors and blowers are considered service equipment, while welders, starters, electrical equipment and the rest of the machines are designated as production equipment. Or a division could exist between Product A-line and B-line, or certain processing stages, whereas maintenance does not separate its data except by department. Whatever division is in effect, both maintenance and accounting should adhere to the same one. The meaning and use of terms such as: production equipment, auxiliary, transport, handling, fixed, mobile, etc., should be equalised.

Similar problems arise when distinctions are made between permanent and temporary, direct and indirect labour. These distinctions may be required to satisfy certain accounting procedures. In some instances union practices may also confuse the issue by insisting on certain craft definitions. In some cases the accounting of costs for handling equipment which serves different departments is often fragmented in an attempt to get clearly defined departmental cost. Such a problem often becomes a red herring confusing issues and causing tempers to rise. Calendar years, financial years, calendar months and weeks are additional hazards since they do not neatly coincide with working weeks. Whether to cut off a month on Fridays or on the last day of the month can be debated. But a uniform decision is vital. In maintenance we, of course, will advocate Fridays, Saturdays or Sunday nights. But sales figures, rent, power consumption are accounted for by the calendar. Unless a consistent procedure is adopted corresponding figures become incompatible, and such problems often prevent us from using the assembled data. In particular the allocation of fixed and semi-fixed costs give rise to difficulties when the end of a month falls in the middle of a week.

Since maintenance is usually poorly equipped with clerical staff anyway, it is suggested that this should not be imposed on them. Cost computation inside the maintenance department should be minimal. Again, when stores clerks are expected to do calculations we are asking for trouble. To make such a system workable the figures available there must be kept up to date and information must be provided to the stores without delay. And then, when it comes to the real issue, the figures are only a duplicate of what is available first hand in the accounting office. Providing the data is not always easy if, for instance, suppliers are not particularly prompt in sending their invoices. As part of establishing the cost of a purchased item, on-cost charges have to be added to cover handling, delivery, insurance, freight, depreciation, etc. Invoices from subcontractors present a similar difficulty and may delay summaries that involve a cut-off date. Therefore accounting figures used by the maintenance department can only be secondary at best and rarely up to date. (These remarks relate to discussions on pp. 228–230 concerning the impossibility of costing jobs with the information available in the maintenance department.) It is often for these reasons that monthly accounting reports cannot be available by the fifth of every month. It is suggested that details outstanding on the twenty-fifth of a month, should be omitted from the current month and included in next month's report, rather than delaying the over-all information.

Now when it comes to calculating the cost of losses due to scrap, second-grade products, rework, delays and down-time we are entirely at the mercy of accounting procedures. We may want these figures to assess the effects of maintenance on these cost components. It is doubtful whether many companies maintain regular computation on these costs or whether they are made at all. It is usually done at the special request of an executive who intends to prove his point in a debate. Admittedly, there are cases where down-time is a sensitive figure and then it is rigidly scrutinised, *e.g.* a steel rolling mill or a paper mill. In these cases, at least, down-time and the losses incurred by delays are accurate. In haulage and earth-moving equipment down-time requires very clear definition to avoid overlapping with unoccupied time, *i.e.* unscheduled shift time. Similar problems, of course, exist with most plant equipment.

For maintenance control purposes it is best to use costing figures only where unavoidable or when they result from straightforward bookkeeping, *e.g.* invoiced purchase or subcontracts. The total maintenance payroll is a case in point as long as payroll only is being considered. This is simple to compile at any level. The cost of a productive hour is a different case. A productive maintenance hour is that which is directly chargeable to cost centres. Deducting ineffective and indirect hours from the departmental total, we obtain the productive hours in a different way. When total payroll is divided by this figure the cost of a productive hour is obtained. It is certainly a revealing figure and a very indicative one. However, the way in which it is calculated offers a wide range of pitfalls. Unless a fixed procedure, defined and traceable, is followed, an occasional calculation may be misleading, as is often the case.

Much of the difficulty lies in the fact that Accounting keeps the figures and adds the on-cost rates over which they have exclusive rights. Maintenance does not as a rule have a say in the calculation, the attitude being, "Who are you to tell me how to calculate my figures?" The solution lies in an itemised procedure that uses figures which emerge from direct recording, *e.g.* payroll, tool costs, materials, power and other items. If a distinct procedure is not followed, figures become "fogged" and maintenance managers will have a hard time convincing anybody that due to their efficiency and efforts the cost of maintenance to the company has been reduced. Figure 79 shows some of the control data as they relate to maintenance, production and the plant.

Closer breakdown of hours and costs as shown in this chart can serve as a guide to the introduction of a recording system. As can be seen from the table the original data can be classified in many ways. This depends partly on whether we charge costs and hours to cost centres or whether we want to know the percentages of different groups. Knowing the results either way may enable us to draw proper conclusions or take corrective action.

As mentioned earlier, most of this data is readily available without any additional clerical work, provided we have the basic essentials of a work order, a foreman's daily report of hours, a production report and stores issue slips. As the size of the maintenance service increases more procedures can be implemented; however, the basic ones mentioned are indispensable. Under average conditions the paperwork that will carry most of the information is represented in Fig. 80.

The exercise of controls assumes that the figures obtained can be compared to certain benchmarks or indicators. Budgets serve as guidelines against which actual results can be compared. Down-time figures, maintenance manpower and others serve as similar guides. The term "standard" can only be used here in a very restricted sense. There is no industry-wide standard for down-time in, for instance, shoe production or concrete block manufacture. However, our performance for the last six months (if properly computed) can serve as *our* standard against which the results of the next six months can be measured.

If we apply controls for a given period we may assume the figures to represent standards. This is not very accurate. It could be misleading to refer to figures that have piled up in the past without much discrimination as "performance standards." They may include a multitude of foreign factors and components. Therefore, past performance figures can only serve as standards if their origin is quite accountable. In most cases it is advisable to adopt "engineered standards," namely those that contain no irrelevancies. This discussion emphasises the fact that a competitor's figures cannot serve as our guide. Under no circumstances can we compare their figures with ours. No two firms can be identical in every respect so as not to differ in some aspects, such as age of equipment, workers' skill, supervision (quality and density), management abilities, conditions of plant usage, etc. Industry-wide averages are like pictures in a gallery—nice to look at. And by the time such averages become available, they are

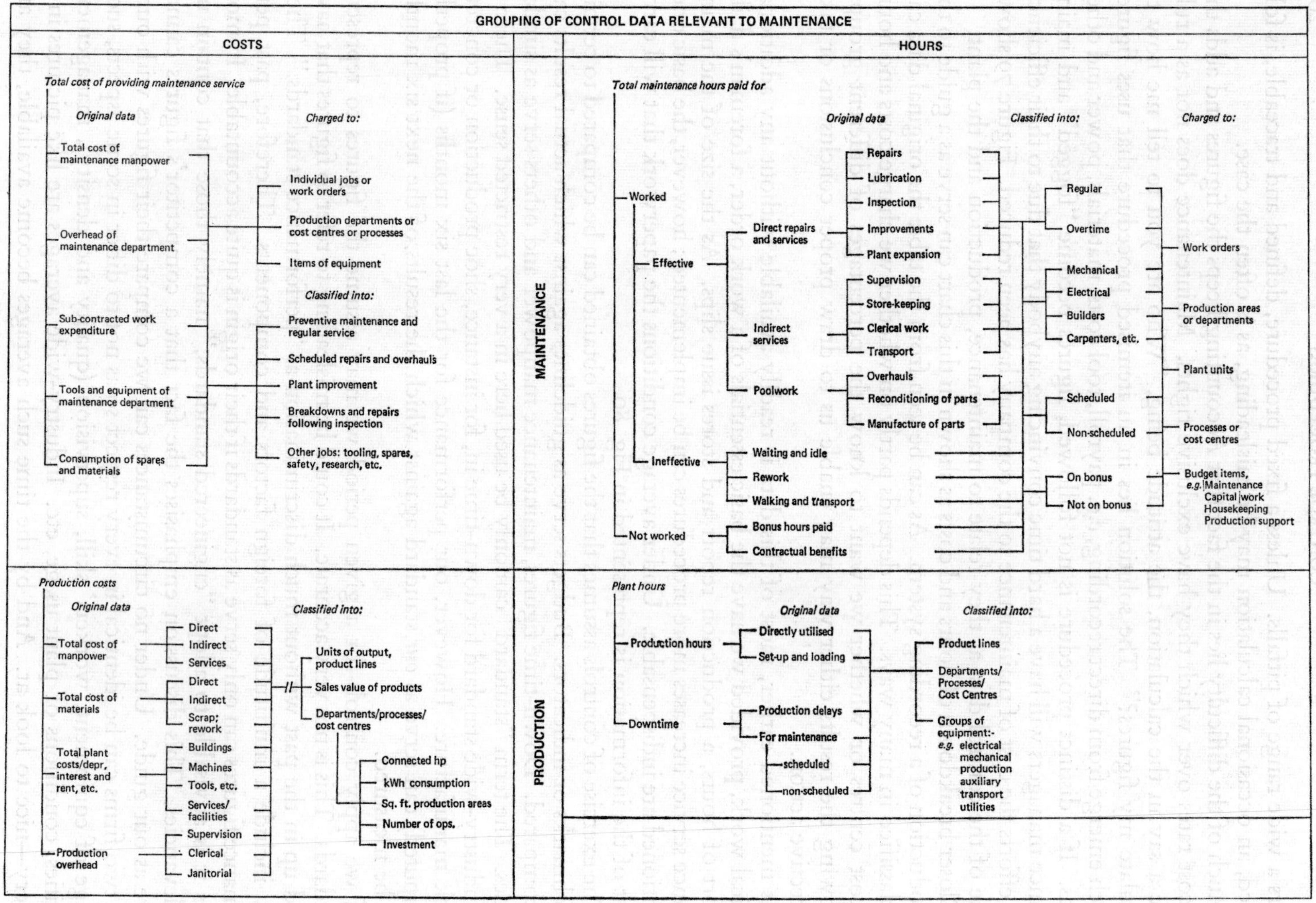

Fig. 79. The grouping of control data relevant to maintenance.

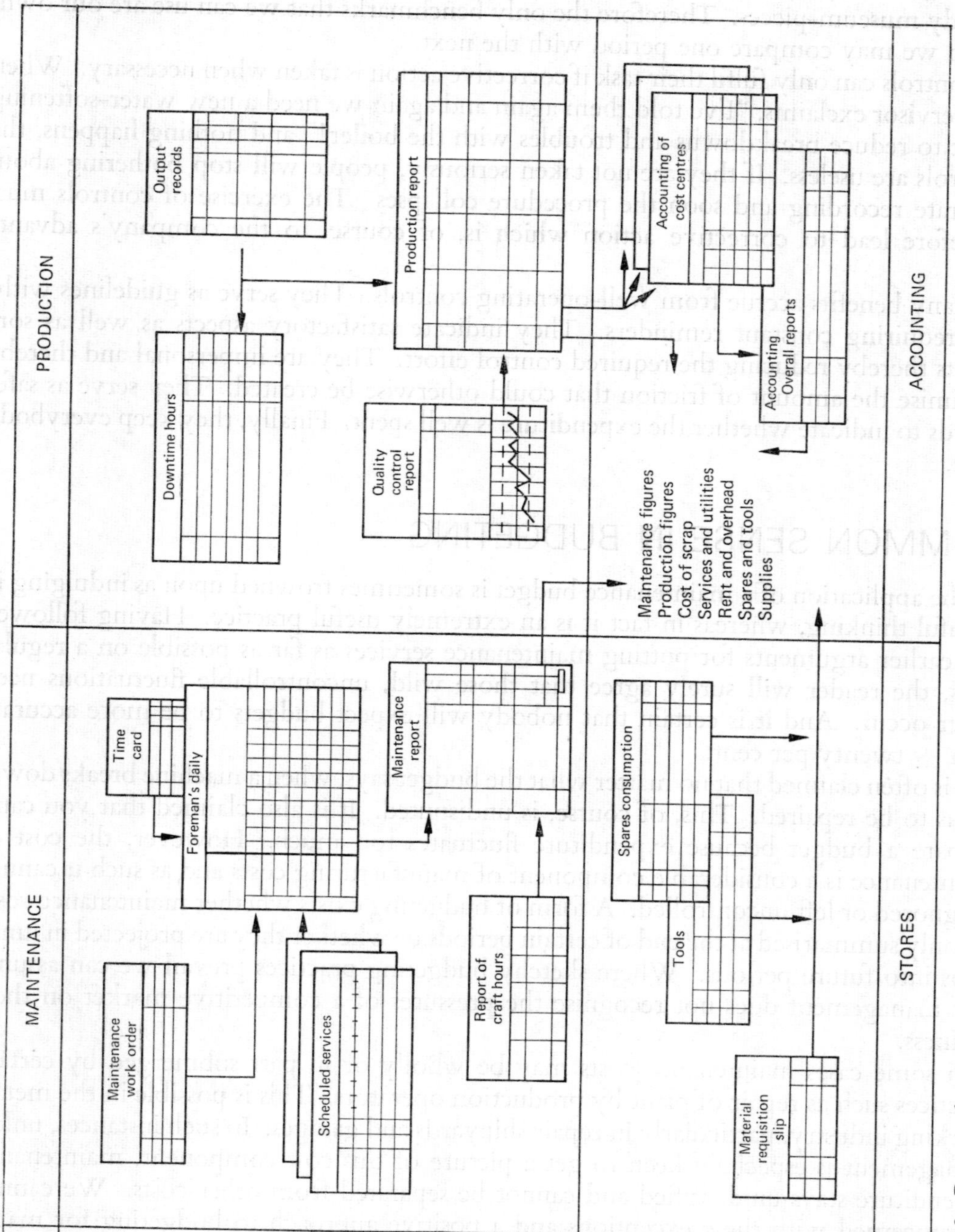

Fig. 80. Paperwork carrying control information in average conditions.

already museum-pieces. Therefore the only benchmarks that we can use are our own, when we may compare one period with the next.

Controls can only fulfil their task if corrective action is taken when necessary. When a supervisor exclaims, "I've told them again and again we need a new water-softening plant to reduce breakdowns and troubles with the boiler!" and nothing happens, the controls are useless. If they are not taken seriously, people will stop bothering about accurate recording and soon the procedure collapses. The exercise of controls must therefore lead to corrective action which is, of course, to the company's advantage.

Many benefits accrue from well-operating controls. They serve as guidelines without requiring constant reminders. They indicate satisfactory aspects as well as sore points thereby reducing the required control effort. They are impersonal and thereby minimise the amount of friction that could otherwise be created. They serve as safeguards to indicate whether the expenditure is well spent. Finally, they keep everybody alert.

COMMON SENSE IN BUDGETING

The application of a maintenance budget is sometimes frowned upon as indulging in wishful thinking, whereas in fact it is an extremely useful practice. Having followed our earlier arguments for putting maintenance services as far as possible on a regular basis, the reader will surely agree that those wild, uncontrollable fluctuations need never occur. And it is certain that nobody will expect budgets to be more accurate than ± twenty per cent.

It is often claimed that no matter what the budget says, when a machine breaks down, it has to be repaired. This, of course, is undisputed. It is also claimed that you can't prepare a budget because expenditure fluctuates too much. However, the cost of maintenance is a considerable component of manufacturing costs and as such it cannot be ignored or left uncontrolled. A form of budgeting exists whether maintenance costs are only summarised at the end of certain periods or whether they are projected in lump sums into future periods. Where sketchy budgeting practices prevail we can assume that management does not recognise the pressures of a competitive market on their business.

In some cases maintenance costs may be wholly or in part submerged by certain practices such as repair of plant by production operators. This is possible in the metal-working industry, particularly in repair shipyards and garages. In such instances, unless management is especially keen to get a picture of this cost component, maintenance expenditure stays unidentified and cannot be separated from other costs. We cannot be concerned with these exceptions and a positive approach to budgeting for maintenance is here advocated.

There are many advantages in establishing a maintenance budget:

(*a*) It allows management to plan its expenditure for further periods.

(*b*) It provides a framework for a cost-conscious operation of maintenance.
(*c*) It enables the exercise of controls.
(*d*) It leads to a thorough scrutiny of proposed operations.
(*e*) It emphasises the need for defining authority and responsibility for the implementation of the budget at various levels of the organisation.
(*f*) It acts as a guide for the maintenance executive while giving him a well-defined degree of freedom to act.

The mere exercise involved in establishing a budget is beneficial. Among other things, maintenance managers are thereby forced into planning ahead or preparing convincing cases for their pet ideas. Proposed items of expenditure must be foreseen, such as new tools, replacement of worn tools, the cost of maintenance paperwork, improvement of central shops and more. There is no doubt whatsoever that "sweating it out" is worth while. When the budget is finished and accepted, this function has a new lease of life at least for the coming year. When all the issues have been thrashed out with management and production the air is cleared of many areas of dispute. It is often a disappointment to have a budget accepted without an argument. It looks fishy. "Could it be a typing error in the summary table or did we estimate too low? Well, it's all over now!"

The factors that dictate the amount to be budgeted for maintenance reflect: (*i*) the policies of management requiring a specified level of trouble-free running, (*ii*) the actual needs of the plant, (*iii*) the conditions of plant usage and (*iv*) the financial position of the enterprise.

The budget can be established in several ways. It can be guessed, or it can be compared with last year's total expenditure. It can be set as a percentage of total factory cost. In processing industries such as mining or refineries, the maintenance budget is often calculated as a cost component of the production volume, *e.g.* so much per ton or per ten barrels of oil. This relates the allowed maintenance expenditure to plant output. We could similarly allow five pence per refrigerator or £20 per car coming off the assembly line. This, of course, can only be done after much experience and experimentation.

All of these methods are as wide open to argument as they are arbitrary. It is very hard to relate maintenance expenditure for the year to any guidelines except a systematic assessment of all its expected components. The amount to be spent relates to the work that is expected for the year (*see* Fig. 34). This predictable amount of work dictates a certain size of the workforce, a certain rate for the consumption of materials, tools and spare parts. Out-contracted work can also be assessed in advance for building maintenance and road repairs, company vehicles and the like. When this is done, there remains nothing obscure about the structure of an adequate budget. Most of the expenses are anticipated and justified. A budget must also include expected project work and shutdowns and reconditioning of the premises.

The last point allows a distinction to be made between expenditure to *maintain operation* of the existing plant and a "beautification" programme or perhaps good housekeeping. These two items often compete for funds, whereas the limits for each should be laid down in the budget in advance so as to avoid squabbles. In some plants appearance and cleanliness can be pursued for business reasons. It is management's prerogative to determine the emphasis to be given to housekeeping, and this should not normally arouse objections. But it should be defined in advance and clearly allocated for keeping the plant clean or "exceptionally clean." A budget for this purpose could sometimes be part of the sales effort or advertising expenditure and linked to a sales volume. Identifying this type of work for what it really is allows a much clearer scope for decisions on how to spend the money. Some of the overlapping responsibilities can also be clarified, such as the amount of janitorial services, gardening, painting, etc. Thus we avoid inadvertently draining maintenance funds for the sake of work leading to better appearance.

The existing literature on maintenance budgeting shows excessive preoccupation with "responsibility for maintenance costs," "authority for expenditure" and with finding a key to charging the costs to consuming departments. Another feature for lengthy discussion is the choice between fixed, flexible or step-by-step budgets. There is no doubt that in this field the accountants have had the upper hand so far. This has had the unfortunate effect of making the subject inaccessible to many maintenance men. Instead of asking "How much do you need to keep our production going and what are you going to spend the money on?", pressure is usually exerted to identify fixed and variable cost items, budget appropriations and hourly cost rates. The latter is the total budget divided by the expected directly applied maintenance hours. Although useful as a device for allocating expenditure, average cost rates are unfathomable to many maintenance men. Other budget requirements state such items as "fixed and variable overheads," "fixed utility expense," "depreciation" and "corporate expense." In short, the subject has been made unpalatable, impracticable and thereby poorly applied.

Budgeting practices often ignore:

(*a*) The effectiveness of the services expected.
(*b*) The ability to supply the service required with the allocated money.
(*c*) The true demand for maintenance services.

The emphasis of the accounting function has been placed, wrongly as a rule, on finding:

(*a*) A key to relate the budgeted amount to a level of production activity.
(*b*) A justification of the budget in view of past experience.
(*c*) A way in which it can be charged to certain operations.

The last and perhaps greatest shortcoming of many budgets is the fact that the list of expense items is not directly recognisable as the ones we are familiar with. "Repair of compressor in spray booths" becomes "Production utilities" under the heading "Variable costs."

To clear the air the following plain facts must be stated: Maintenance expenditure is incurred by employing labour, buying supplies and by operating out of a workshop. The way money is spent on these items depends on the type and amount of work that is done, which ranges from the oil-can to "logical fault finding" and to installing a new process. The recipients of service are the machines and shops that make up the whole enterprise. The expenditure on repairs is conditioned by the "customers" of the service, namely production functions and the machines they operate. The person who is in charge of maintenance has little control over this major cost component although he does have to authorise them on behalf of the recipients. This is illustrated in Fig. 81.

To allocate cost responsibility to the person authorising the work or the person requesting it, is a misapplication of management principles. To attach the responsibility for certain maintenance work to a department or to a person and thereby support or refute his claim for a bonus on a cost-reduction campaign as is sometimes done is sheer lunacy. If that person is a shop supervisor, will he be entitled to a bonus or a fine if he requests a machine to be lubricated when it needs to be done—regardless of what the service schedule indicates?

This type of reasoning has befuddled the budget thinkers. The recurring theme in all dissertations on this topic is "Cost of services was assigned to individuals possessing authority to make decisions" and "the supervisor in charge of the consuming department was responsible . . . while the maintenance supervisor was responsible for the unit cost of rendering maintenance service." (*Control of Maintenance Costs*, p. 65.)[12] All of which serves only one purpose, namely to improvise a way to allocate the expenditure incurred and to clothe it in respectability.

It is also strange to observe that when routine service schedules are prepared, submitted and accepted, costs should be charged to production and emergency repairs are charged to maintenance. The logic of these procedures is wearing thin when it comes to overhauling a lathe. Since presumably the lathe is serviceable but "not too good" until the time of the overhaul, the approval to have it overhauled comes from the chief maintenance engineer who has been delegated the authority to decide by management. There is, of course, no way to ascertain whether the production supervisor has been running it into the ground, or whether the company has been trying to defer for too long the cost of purchasing a new one, although the recent line of production would have demanded the purchase of a larger heavy-duty machine long ago. In such instances where does the responsibility lie? Who is accountable?

Returning to Fig. 81 it is evident that a cost item can be identified completely by a coding number covering the variables tabled. Here letters are used but with numerals we could establish a decimal coding system. Thus the cost of replacing a bolt–nut–washer set on Pump No. P101 during a repair job in Department A, by fitter No. 34, can be identified as:

Aa F34/Ca/Da/Eb Ap 101/—57 pence,

by following the letters assigned in the table to the various groupings. If this identification looks rather excessive let us list the headings that appear on an IBM Maintenance

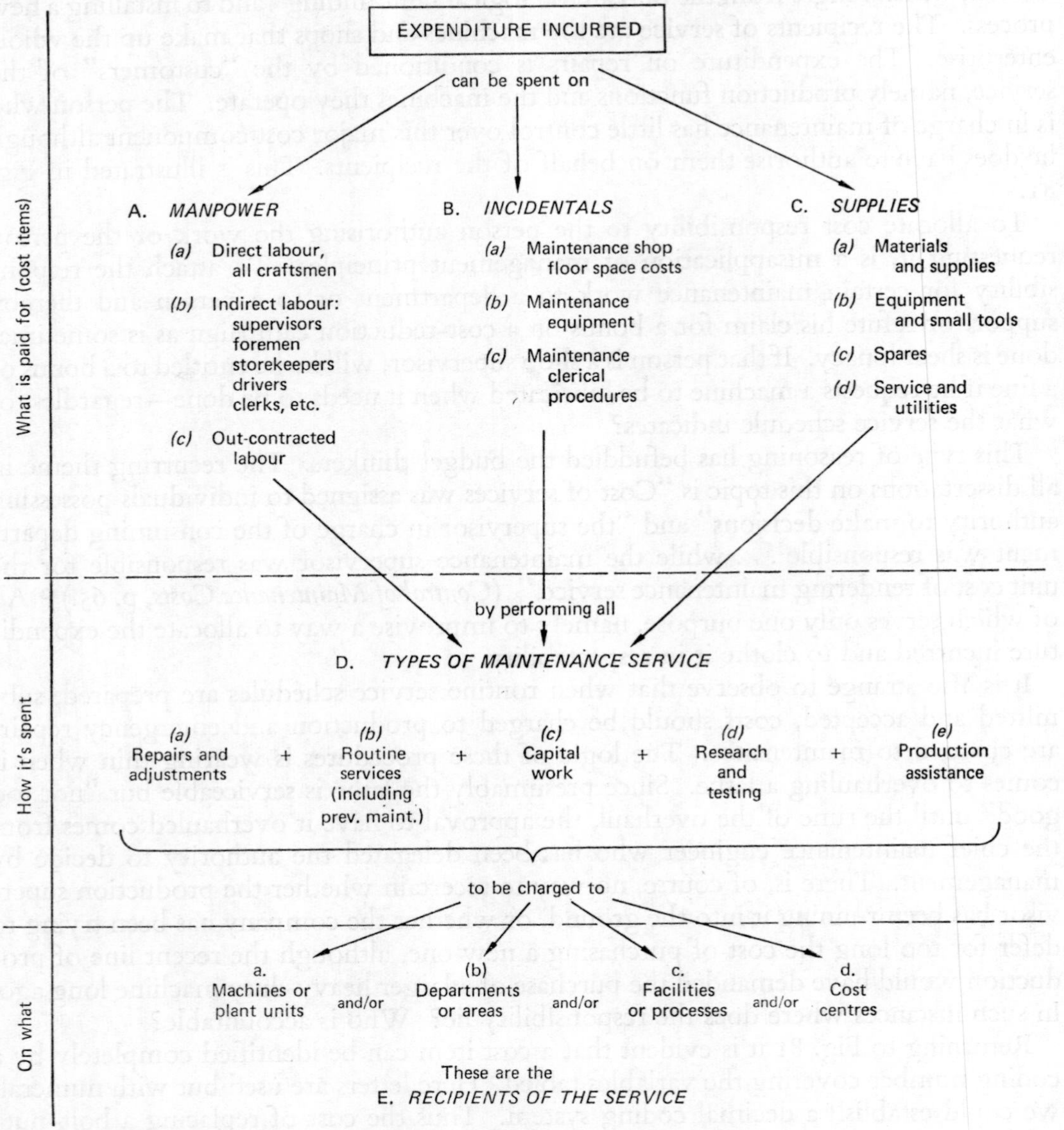

Fig. 81. Schematic flow and classification of maintenance expenditure.

Repair Order 784 103094 KGN 7/65 (quoted from *Mill & Factory*, New York, December 1967):

Maintenance Order Number
Reference Number
Shop Order Number
Appropriation Number
Account Number
Department charged
Serial Number
Identification Number

This list of identifying numbers, conveying no useful data, is by no means exceptional and is only quoted here as an example. In most instances job number, plant number and department will suffice. There is no doubt that the above-mentioned 57 pence have to be accounted for and charged to P101–Department A. Similarly, all work has to be identified, coded and charged to recipients.

However, the budget is built up in a reverse sequence from that appearing in Fig. 81, namely, from the bottom up. First, the services to be provided to all recipients are tabled. Later a price tag will have to be worked out for each. Next, expected expenditure for each plant unit must be assessed. A listing of the available manpower and their assignments will provide the labour cost. The expected consumption of all supplies and the incidentals can also be estimated. To do this in a more systematic way a "three-dimensional budget" form is here proposed (*see* Fig. 82).

This form requires the workload expected in every department to be assessed and broken down into the type of work to be done. This is to be filled out in the right wing of the table. If there are eight departments or cost centres to be charged and twelve types of work that have to be paid for, ninety-six spaces in the table will be filled.

When the horizontal totals on the right are available, the left wing is to be filled in, proceeding from the known to the unknown. The column relating to labour comes first, followed by departmental overhead. The balance is to be judiciously subdivided between materials, tools and spares as expected. The total in this wing has to correspond with the total in the same horizontal line in the other wing. Therefore the two wings show a different breakdown of the same over-all sum.

The lower part of the table allows the inclusion of those oft-debated items. "True" maintenance work is first totalled and the percentages are calculated for all columns. When the lower part containing "beautification" and janitorial work is filled, this procedure can be repeated to show the revised percentages of the over-all totals.

For the purpose of accounting, we assign major code or account numbers to all columns and lines, and whenever costs are incurred we hit upon the right identification. Next year's budget will be so much easier to prepare!

It must be remembered that the suggested itemisation must be such that the items are mutually exclusive, and there should be no overlap. When a job card is prepared and

THREE-DIMENSIONAL BUDGETING								For the year..								
Expenditure by groups of items								Identification			The budget for fixed assets					
Labour		Stores and supplies			O/Head etc.	Totals	%	TYPES OF WORK			Dept. 1	Dept. 2	Utilities	Grounds	Miscell.	Totals £
Direct	Indirect	Materials	Tools	Spares						Accounts no.						
								Repairs	Mechanical							
								Repairs	Electrical							
								Repairs	Bldg. and carpt.							
								Repairs	Paintwork							
								Regular and routine servicing	Mechanical insp.							
								Regular and routine servicing	Electrical insp.							
								Regular and routine servicing	Lubrication							
								Regular and routine servicing	Parts replacement							
								Shutdown and overhauls								
								Plant improvement and capital work								
X	X	X	X	X				Out-contracted work								
								Central maintenance shop								
								Other								
								Sub-totals		X						
%	%	%	%	%	%		100 %	Percentages		X	%	%	%	%	%	
								Housekeeping								
								Safety								
								Janitorial, etc.								
								Overall–totals		X						
%	%	%	%	%	%		100 %	Revised percentages		X	%	%	%	%	%	

Fig. 82. "Three-dimensional" budgeting sheet which allows figures to be checked and summarised in three different ways.

later filled in, no question can arise as to the correct account number and there are at least eight ways in which the same expenditure can be accounted for (*see* Fig. 83). The code system is often pre-printed on job cards next to the group description. Collation of data is thus assured to follow the account numbering and the budget.

By adhering to this procedure we can easily identify direct and indirect costs and the cost of lubricants, consumable tools, departmental paperwork and other costs. Whether to allocate this latter figure to recipient departments by square feet of area, man-hours worked or production value is entirely an accounting problem. Only the totals of these items are of interest to us. This field incidentally, is a prime object of cost-reduction campaigns. Although the expenditure in this group reflects upon the maintenance

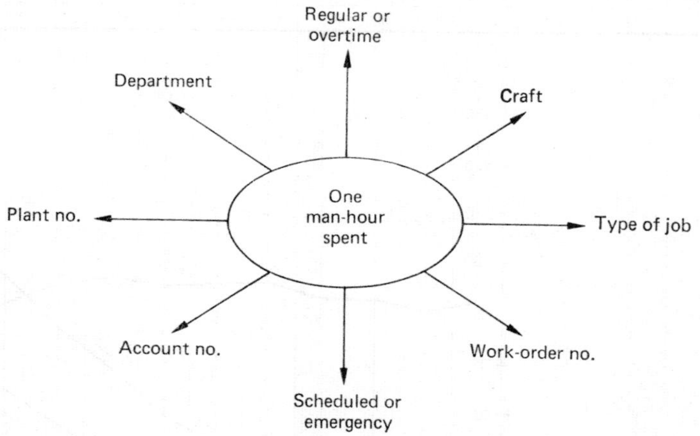

Fig. 83. At least eight ways in which one man-hour can be classified.

manager's control over his department, excessive emphasis could lead to parsimony where more would be lost than gained by excessive trimming of costs.

The practical value of this kind of budgeting for control purposes can easily be demonstrated. A one-twelfth of every figure in the table is the monthly budgeted sum. By plotting the values on Z-charts we can visually follow progress throughout the year. Otherwise we could work in percentages. If, for instance, direct labour made up 48 per cent of the total budget which was equal to £100,000 for the year, the *monthly* direct labour payroll should not exceed 4 per cent of the yearly budget for that item, *i.e.* £4,000 being one-twelfth of £48,000.

When the cost of work done is analysed monthly into repairs, regular services, overhauls, etc., the percentages can again serve as performance guides when compared to the annual figures appearing in the proposed form. At year's end or at semi-annual periods variances between the proposed and actual percentages would have to be investigated. It could either be due to a poor projection for the year, or a non-performance of planned work.

Staying within the projected budget or exceeding it, hinges firstly upon the accounting computations, secondly on reliable recording of work on the shop floor and thirdly on outside fluctuations of costs. It becomes evident that if we were to budget in hours instead of money, a greater degree of conformance could be reached. That is the purpose of operational controls which deal in units *other* than money. These are discussed in a subsequent section.

The use of Z-charts for the main items of expenditure is to be recommended. The chart derives its name from the shape of the resulting curves (Fig. 84).

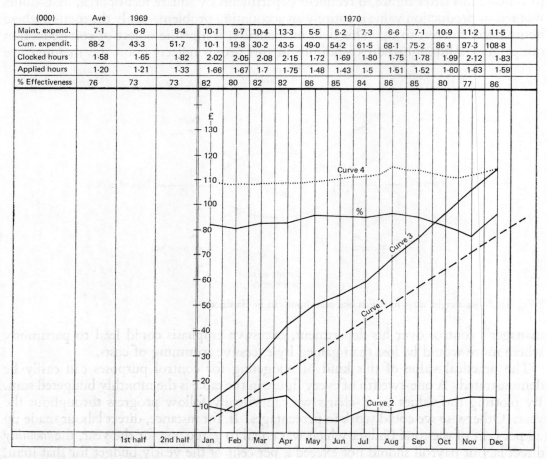

(000)	Ave	1969		1970											
Maint. expend.	7·1	6·9	8·4	10·1	9·7	10·4	13·3	5·5	5·2	7·3	6·6	7·1	10·9	11·2	11·5
Cum. expendit.	88·2	43·3	51·7	10·1	19·8	30·2	43·5	49·0	54·2	61·5	68·1	75·2	86·1	97·3	108·8
Clocked hours	1·58	1·65	1·82	2·02	2·05	2·08	2·15	1·72	1·69	1·80	1·75	1·78	1·99	2·12	1·83
Applied hours	1·20	1·21	1·33	1·66	1·67	1·7	1·75	1·48	1·43	1·5	1·51	1·52	1·60	1·63	1·59
% Effectiveness	76	73	73	82	80	82	82	86	85	84	86	85	80	77	86

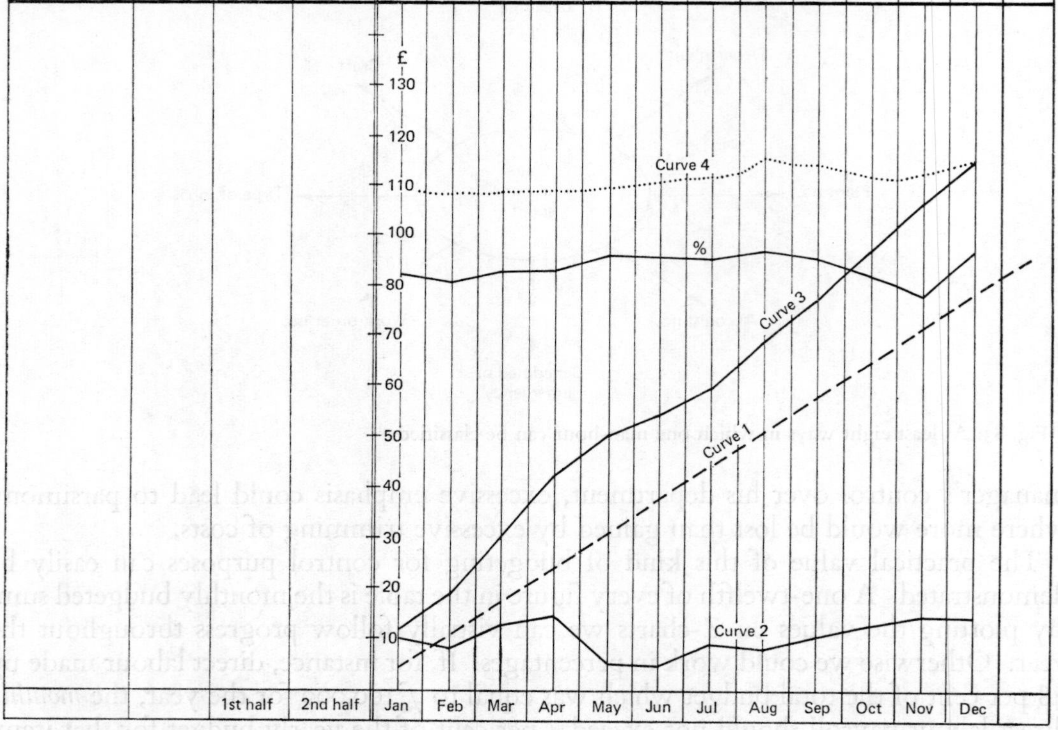

Fig. 84. An example of a Z-chart for cost control.

Curve 1 is the cumulative target line that starts from zero at the beginning of the year to the total sum at the end of the year. When seasonal work is foreseen the line will not be a straight one, however, assuming a year-round operation we can chart it as such.

Curve 2 results from plotting the monthly sums, and it may show either a rise or a fall or a reasonable degree of fluctuations between individual periods.

Curve 3 shows the cumulative sums as we add the February sum to the January figure, and subsequently, March and so on. Under ideal conditions it should follow the target line approximately. If, however, a wide upward gap results by April or May, it could indicate that the year-end sum is most likely to be exceeded, unless corrective measures are soon taken.

Curve 4 is a moving twelve-monthly total. An upward trend or a downward one in this line indicates a departure from last year's expenditure.

The chart should allow actual values to be shown along the same time-scale. These should first be posted by a clerk engaged in obtaining them, and the points can then be plotted by the user of the chart. This division minimises plotting errors.

From the preceding discussion it will by now be apparent that a maintenance budget does not necessarily depend on projected production level of activity such as size of direct (production) workforce, number of shifts, floor-space occupied or units of production. In fact, we may be reducing current production and at the same time expanding our facilities, and this may mean more maintenance work than in previous years. It may also happen that the company decides on a massive plant-improvement programme to catch up on previous years of neglect. There is therefore only one way to justify a proposed budget, "Is this the amount of work we expect to do during the coming year?"

Only a stagnating company, or a slowly declining one, can have a stable maintenance budget year after year. There is another argument that we sometimes encounter. "We cannot afford it." This argument is difficult to refute but so are the detrimental effects of a severely curtailed maintenance budget.

We must also be careful to choose a fixed, an adjustable or a step-by-step budget. Having come so far the reader may have already arrived at his own conclusions in this matter. It is obvious that some of the expenditure rises in direct proportion to volume of production, while some items will rise during off-season or transitional periods.

Applying a fixed budget for the year for all items with one-twelfth to be spent every month is not more than guesswork, and applying an adjustable budget that increases with the volume of production is arbitrary. Maintenance expenditure will indeed rise with more workers employed, more shifts worked and more units produced. The converse is not as obvious. But the use of figures according to these variables can be merely a guide. The control figures described in the following section may indeed contribute to this approach by indicating *levels of expenditure*, or so-called *rates*; *i.e.* *X* per man-hour, *Y* per square foot, etc. A variation in activity will automatically indicate an increase or decrease in the level of expenditure. However, budgets cannot be set by these rates, they can only be used to explain the variations in expenditure. Temporary fluctuations of these figures are also not conclusive. Budgets should only be revised if the production activity is to continue at a different level for a long period.

The system described in this book has proceeded through the stages that would ensure controls to be effective, namely to reflect variations according to demand in service. A continuous check on the *workload* will indicate the demand and data from the shop-floor will show the *rate of implementation*. These two variations must be reconciled. As the *demand for service* fluctuates, the budget can be adjusted. But to adjust the budget without considering the maintenance activity is not advisable.

OPERATIONAL CONTROLS

This term refers to control which is expressed in non-monetary units. The term "control," as stated earlier, implies (*i*) availability of data, (*ii*) comparison of targets or bench-marks, (*iii*) identifying excessive deviations and (*iv*) ability to take corrective action. It may also be specified that controls should indicate the area where action is necessary. However, this can only be done indirectly by correct interpretation. The assumption is that a poor result as compared to a certain target would point to the necessity for certain action so as to improve that particular result. This is rather too much to expect the control data to do; while it can indicate it cannot propose solutions. Furthermore, there may be more than one kind of action that could lead to the same result. For an indicator to prove conclusively, for instance, that one additional foreman is to be appointed, is (*i*) to expect it to sense that one is required, (*ii*) that he is available and (*iii*) that there will be a need for him for a prolonged period. Naturally, were there any proven deficiencies in supervision as shown by the controls, we could either train and improve the existing foremen or decentralise a number of men, or take other measures that would improve the effectiveness of supervision in the department. There is, therefore, no way that a sensor can prescribe corrective action. For this purpose we apply human judgment.

Performance measurement relates to the expression of results in figures that evaluate a certain activity in relation to targets. Where "control" tells us *what* is happening "performance measurement" shows *how well* or *how badly* it is being done. Control is an indicative tool of management and provides a short-term view while measurement is an assessment and, being more cumbersome, can be effective only on a more long-term basis. Both techniques serve the same purpose, namely to act as a guide in fulfilling the objectives of maintenance service within the activities of the enterprise. The payment of incentives to maintenance staff on the basis of these results *is* possible. However, a distinction should be made between figures that are applied to assess "shop management" and those applied to the over-all conduct of maintenance functions within the broad company objectives. Controls may serve as a basis for incentives for "front-line" management, while performance measurement can be used as a basis for assessing the effectiveness of middle and top management levels. This is dealt with in the next section.

As we have seen in the previous section, there is an abundance of data generated at the point where work is performed. The mere tightening of a belt on a power drive, if amply recorded, would identify the belt and its location in five details, the person performing it in three details and the data on performance in two. These figures and their summaries are primary data, as recorded at first hand. Secondary data are combinations and reclassifications of primary data. Tertiary data have undergone further processing. As an example, total "direct electrical maintenance hours" is primary data, "percentage of hours on planned electrical service" is secondary and "cost of electrical breakdowns" is tertiary.

As can be seen, each further level of calculation requires additional information. The electrical maintenance foreman disposes of the primary data (if he has the procedures for it) and he *can* compile the secondary level. But unless he is kept informed by accounting he cannot reach the third level. It is also questionable whether being in possession of that information will do him any good. After all, he only needs to know the total of electrical breakdown hours to realise whether the situation is good or bad. Whatever the results, he can't do much about them, can he? Top management, on the other hand, seeing the total cost of this item, can decide what corrective action to take. Who knows, perhaps they will fire the electrical foreman. Looking at information in this way is conducive to providing the right data at the right level.

For controls to be feasible the following prerequisites are required:

(*a*) A proper recording procedure.
(*b*) Accurate recording of specified data.
(*c*) Transmission of data to destination.
(*d*) Compilation and appraisal of data.
(*e*) Presentation of data for appropriate action.

Assuming the first three to be satisfied, the last two conditions are often stumbling-blocks. How and what to compile? How to summarise and tabulate? How to present so as to lead to appropriate action? It must be realised that a certain piece of information has more than one context (or co-ordinate) and could be grouped in several different ways (for instance, *see* Fig. 84).

Rewriting the same information manually under different headings can be very laborious, costly and sometimes impracticable. If done by EDP it is less so. The same input is rerun in several ways so that we get a print-out arranged by Work Order No., by sequence of consuming departments, by crafts, etc. This is the reason why a multi-dimensional recording is strongly advocated in favour of simple summaries. To illustrate, a single-parameter versus a multi-parameter tabulation is shown in Fig. 85.

Referring back to Chapter 5, p. 67, Fig. 22a, we note the remarks concerning pre-printing of classifications on job cards, to allow the appropriate ones to be checked. While extracting the data and posting it, a multiple selection has to be done. Which trade? Was it regular or overtime? How much is the total? What department? In Fig. 85, emergency jobs can be prefixed with the letter "E" (E W/o) and another dimension is added to the table. By presenting data in this form we obviate the necessity for asking for three to four additional clarifications.

Now we come to the crux of the problem. What do we want to control? What activities do we want to keep an eye on? What do we intend to do about the results when we get them? Evidently, just getting the figures is of no use unless we employ them to some purpose. The following information can be used at shop-floor level and in a consolidated form at management level:

(*a*) Application of manpower (time utilisation).
(*b*) Work completed and outstanding.
(*c*) Performance in terms of cost.

MAINTENANCE HOURS BY DEPARTMENTS							
Dept.1	*Hrs*	*Dept.2*	*Hrs*	*Warehouse*	*Hrs*	*Powerhouse*	*Hrs*
Work order No. 123	15	Work order No. 234	3½	Work order No. 245	22	Work order No. 456 etc.	7

ANALYSIS OF DIRECT MAINTENANCE HOURS													
		Craft hours								*By departments*			
	Work orders	*Mech.*		*Electr.*		*Bldg.*		*Total*		*Dept. 1*	*Dept. 2*	*W/H*	*P/H*
		Reg.	*O/T*	*Reg.*	*O/T*	*Reg.*	*O/T*	*Reg.*	*O/T*				
Repairs	Work order No. 123	3	2	3	2	5	–	11	4	15			
	Work order No. 231			5							5		
	Work order No. 345	14	4	4				18	4			22	
	Work order No. 456					7½		7½					7½
PM Schedules	PMS No. 15	40						40		40			
	PMS No. 34		38					38		38			
	PMS No. 3a	40						40			40		
Cap.	C-W No. 51												
	Regular hrs	94	X	50	X	12½	X	159½	X	X	X	X	X
	Overtime hrs	X	6	X	2	X	–	X	8	X	X	X	X
	Total	100		52		12½		167½		93	45	22	7½

Fig. 85. Single-parameter versus multi-parameter tabulation of hours.

Again it is emphasised that at this stage we work with *primary* data which has to be arranged so as to allow a meaningful interpretation. In the chain of command every level has to contribute its part to ensure this outcome. Workers have to report promptly and correctly. Requestors have to identify plant number and department. The foreman records date and time taken and checks satisfactory completion. The time clerk has to collate data properly and the craft supervisor scans and checks the data. Exceptional figures and variances need to be verified, rechecked and investigated. Points of special interest in the tables are to be transmitted to the next higher level with remarks for appropriate action. Controls that do not undergo this chain of sequence will be useless.

Time is the prime commodity of every maintenance department. It is indisputable that the *use of time by craftsmen* and foremen alike is of paramount importance in maintenance. If their time is well occupied, work gets done and jobs are completed sooner so others can be started. When many jobs are in progress the foreman is kept busy making technical decisions, giving instructions, and in general being gainfully occupied. Whether he is the person generating such a situation or whether the workload is enough to make workers rush to their jobs, is not the question here. This particular issue can be relevant to the payment of incentive bonus. The information we seek here is whether requests are dealt with promptly, how far workers are using their time on actual work and whether the work is effectively performed.

Therefore what we have to show for control purposes is the *total time that was clocked*, *i.e.* available and paid for, and the *way the hours were used*. Hours that are paid for, arising out of contractual provisions, *e.g.* sick-leave, vacation, holidays, etc., are not of prime interest, although a general picture of absences are mandatory from the personnel management point of view. Thus, within *each craft* the hours "at-work" are to be presented as shown in Fig. 86.

When a time-based bonus scheme is in operation, summaries can show a division between jobs performed against time estimates, namely bonus "coverage" and time on day-rate, *i.e.* not estimated. In such instances total time saved against the standards (allowed times) would also be tabled (the percentage arrived at belongs to the discussion dealing with performance measurement). In instances where time estimates are made for purposes other than incentives, such as scheduling control of overtime or input of hours, similar calculations can be made, *i.e.* planned versus actual.

Next we deal with the work order situation. (This is synonymous with job card, work ticket, work request, MRO (maintenance repair order), etc.) We assume that emergency jobs and requests for repairs are handled promptly and we want to know what their frequency was and where they occurred.

This group of work orders refers to irregular jobs of all descriptions. Plant improvement and repair jobs are here included as well as manufacture (or fabrication) of replacement parts. Some parts of project work and some individual overhauls could also be included in this group.

To arrive at the delay rate of work orders we have to keep track of all orders that are being processed. All orders on-hand at the end of the week would be counted and

MAINTENANCE HOURS UTILISATION					
Description	*Regular hours*	*Overtime hours*	*Total*	*On-bonus*	*N.o.B.*
1 Effective time on jobs:					
Fixed assignments					
Emergencies					
Regular repairs					
Plant improvement					
Schedules, Routines					
Total					
2 Ineffective time:					
Waiting, idle					
Rework					
Transportation					
Total					
3 Indirect hours:					
Supervision					
Clerical					
Training					
Stores					
Transport					
Total					
Total					

Fig. 86. Table for calculating the utilisation of maintenance manpower.

grouped into those received in the current week, those due since last week and orders two, three and four weeks overdue. By multiplying the number of orders in each group with the amount of delay that they represent, *i.e.* the number of weeks that they have been pending, we obtain a total figure of week-delays. This is shown in Fig. 87. A rising total of week-delays would indicate whether the backlog of work is getting excessive. This, of course, would also take care of complaints of long delays often made by requestors. Some companies attach great importance to this particular control measure.

Now we would like to know what type of jobs were handled and how many there were in each category. The groups that can be separated are:

Repair jobs.
Plant improvement.
Parts manufacture.
Project work, subdivided into partial jobs.
Schedules and regular services.

Plant improvement jobs (also called development work, investment, capital work, appropriation, expansion, modification, alterations, transfers, etc.) as well as repair jobs vary from small adjustments to large-scale project work. Ways must be found to post their completion. Perhaps the best approach is to group them as *minor, small, medium* and *large* jobs which would be of different lengths, less than one-half a man-day duration, less than one man-day, less than ten man-days, and under thirty man-days. Jobs in excess of these would be measured in estimated project weeks. Overhaul of machines could fall into this group or could be considered as repairs, but plant shut-downs or installation of new plant would be project work.

As regards project work the following remarks are applicable. Major projects should be broken down into separate goals in their required sequence. Naturally, the individual teams will work on separate work orders and a distinction will also be made between crafts. A large project can thus become manageable with partial orders being processed as their time arrives. When the detailed jobs of a project do not extend over many weeks, the control of completion can be in man-weeks or man-days instead of project weeks.

The more difficult part of reporting completed work will be the scheduled services, such as lubrication, inspections, replacements, etc. These should all be coded as to daily and weekly content and location and time required for completion. Thus L55/A2/420 will denote lubrication schedule number 55 in department A.2, time allowed 420 minutes.

CONTROL CHART FOR WORK-ORDER TURNOVER

	Week ending on date:	Jan.5	Jan.12	Jan.19	Jan.26	Feb.2	Feb.9	Feb.16	Feb.23	Mar.2	Mar.9
a	Total no. of incomplete work orders on hand	19	24	20	20	24	22	29	30	31	etc.
b	2 weeks overdue	11	11	9	7	8	7	9	10	10	
c	3 weeks overdue	5	7	6	9	11	9	16	15	17	
d	4 weeks overdue	3	6	5	6	5	6	4	5	4	
e	Total week-delays	49	67	56	65	69	65	82	85	91	
f	Ave. forward load wks	1	1	1	1	1	1	1½	1½	1½	

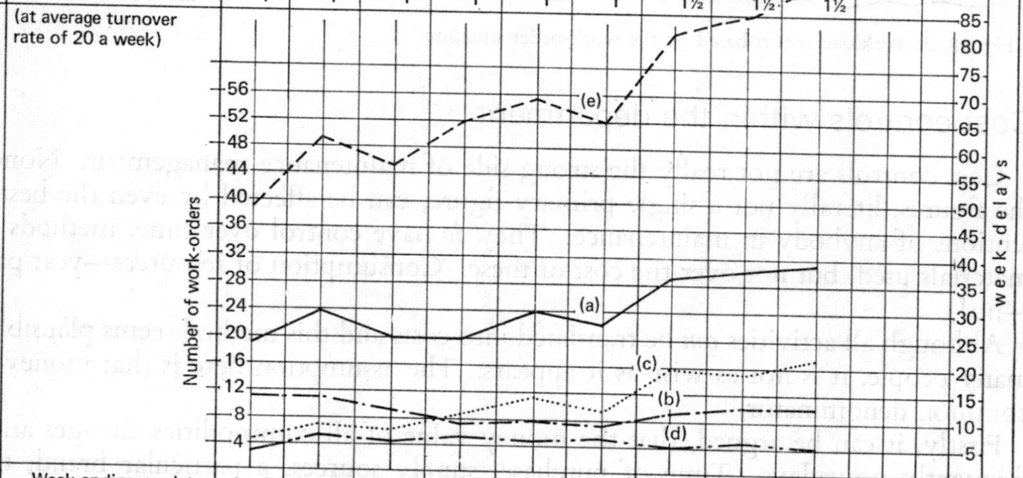

Fig. 87. Control chart showing weekly turnover of work orders.

Keeping a weekly check on the workload completed weekly and adding the newly received orders can be implemented on a sheet as shown in Fig. 88.

To implement this type of control, clear-cut procedures would have to be faithfully followed by the planning clerk. For instance, all work orders would have to be channelled through him after approval. All schedules relating to routine, planned and preventive services would have to be scanned weekly and the allowed times duly extracted and a close check has to be kept on their completion. Although this job is not easy it is not beyond average capabilities.

NUMBER OF WORK ORDERS HANDLED								Period ______
	Stage of progress / *Type of jobs*	No. of orders brought forward from last week	Received this week	Total workload	Completed this week	Balance to be carried forward	Representing an estimated number of man-days	Group totals
1	Repair and plant improvement: man-days Minor (up to ¼ man-day) Small (up to 1 man-day) Medium (up to 10 man-days) Large (up to 30 man-days) Emergencies (all sizes)							
2	Non-project overhauls (weekly) planned weeks							
3	Project work (planned weeks)							
4	Regular and routine schedules: no. of schedules Lubrication schedules PM service schedules PM inspections PM replacements							

Fig. 88. A workload report based on the work-order situation.

Cost controls within the department

Cost controls are not really the strong side of maintenance management. None of the figures, literally not a single primary figure, can be affected by even the best intentions of anybody in maintenance. They *do* have control over time, methods and materials used, but not over the cost of these. Consumption of resources—yes; prices —no!

Although all activities *can* be translated into costs and this method seems plausible to many people, it is not as wise as it appears. The assumption here is that money is a common denominator.

Firstly, it can be argued, that the money value of all commodities changes all too frequently nowadays. Time of purchase, supply sources, a particular brand, these factors have a strong effect on total costs of any one item. Secondly, the total cost of replacing a bearing can be identical in two instances even though circumstances could have been, or were, altogether different. The major cost in one could have been the labour cost on Saturday night, whereas in the other instance, it would be air-freight charges from Timbuktu. Are the high costs of these occurrences to reflect badly on maintenance or are we to conclude that two instances provide statistical proof about the cost of such repairs? Thirdly, in providing maintenance services you have to do your best with what is available. Sometimes costly materials have to be sacrificed because none other are available. Again, what does a cost summary prove? Fourthly, when primary resources (labour and materials) are scarce and controlled, there is little purpose in recalculating the same figures into costs. Granted the budget is in terms of money and the feedback to it must be made in the same terms, the fact remains that at the operational level within the department these cost figures are not very significant, since you cannot do much about them anyway.

In some companies the maintenance department is expected to calculate its own costs. However, the maintenance department *does not* pay the wages, buys neither the spares nor the materials and neither does it pay for rent, insurance or power. Therefore, unless an extension of the accounting function is delegated to the maintenance department, they only "*know*" about the costs *if* they are being kept informed. No other department of the enterprise except perhaps research and development has to know about its budget and its performance. There are no accountants in the tool and die section, in the transportation department or in quality control! It is therefore a puzzle how the maintenance function could ever be expected to cost each stores issue and each job card. The only cost data that the maintenance department could be in possession of is the hourly rate of crafts (without administrative overheads) and the cost of whatever is consumed *if*, and a great if at that, the price-tag of items were always available at the stores, which is far from being the rule. The accurate pricing of stores issues, particularly spares and tools, presents many obstacles. Supplies are received in many ways and forms and are seldom accompanied by final invoices. Therefore in most cases the prices cannot be promptly posted on stock cards. Quantities issued are not always final and sometimes identification of items is deficient. Much of this prevents early and complete costing of stores issues. Therefore to impose cost control and expenditure management upon maintenance will both require additional manpower and will constitute a duplication of work done in the central accounting department. Therefore the notion of exercising expenditure controls in the maintenance department is but wishful thinking.

One result of trying to calculate costs in the maintenance department is that a heap of paperwork is kept in "pending" files awaiting one or another detail of cost information. Thus postings and period summaries are delayed.

Thus, cost controls as such are not as practical in our field as to justify their wide support and promotion. They are not within the sphere of influence of either supervisors or craftsmen and the cost computation within the department is often a farce and always burdensome and costly. Nevertheless, let us try to make a brave effort to propose a feasible compromise.

The cost components that maintenance *is* able to control to some extent are payroll and its subdivisions, *consumption* of tools and supplies and issue of spares. Indirect costs tend to be stable and their computation can serve no useful purpose. The exception to this may be in the employment of handling service that can be billed to maintenance by another department or subcontractors. The cost of out-contracted work or overtime again come within the doubtful field. Neither can really be controlled, only their amount can be summarised. Overtime should be justifiable. It should not be dispensed magnanimously but it often *is* essential. Thus a tabled cost summary can be as shown in Fig. 89.

MAINTENANCE EXPENDITURE SUMMARY

	Maintenance sections / *Cost component*	Central maintenance office	Mechanical section	Electrical section	Building section	Carpenters section	Painters section	Totals
Acct. nos.								
	Payroll:							
	Direct regular hrs							
	Direct overtime							
	Indirect total							
	Tools and supplies							
	Spares issued							
	Contract work							
	Totals							

Fig. 89. Summary of current maintenance expenditure.

Tools and supplies as well as spares have to be charged with an added overhead cost which covers ordering, stock-keeping, storage, handling, depreciation, obsolescence and interest on invested capital. These can be added in the form of a percentage factor deduced from calculating their value in previous periods.

It is interesting to note that maintenance *creates* value at least in two tangible ways, namely (*i*) the manufacture of spares or fabricated replacement parts which otherwise would have to be purchased and (*ii*) revaluation of plant after overhauls. These values could be included among the control data if they occur regularly. Developing this line of thought a little further, one could surmise that the total *commercial value* of maintenance services could often exceed the budgeted expenditure. A price-tag on every repair, overhaul and lubricated service would undoubtedly be an eye-opener to one and all. This approach is, of course, the basis for outcontracting maintenance. Whether that arrangement is totally advantageous is yet to be proven.

PERFORMANCE MEASUREMENT

At this late stage it may be rather provocative to ask: "Who says we can measure maintenance performance and anyway why do we need to measure it? And again, what are we to measure, efficiency or effectiveness?" Efficiency is a ratio of utilisation wholly expressed in figures. Effectiveness is a measure of the effects that cannot be converted into unitary data but can nevertheless be assessed.

In sharp contrast to production work, the performance of maintenance activities does not lend itself to expression in simple figures. In production, units of output and cost per unit (in terms of all input of resources) are enough to assess how well or how badly the results are achieved. Again, in production, we try to *produce as much as possible* as *economically as possible*. In maintenance, this simple relationship cannot be used to assess performance. Measurement requires the expression of results in figures. Again, measurement is not done for its own sake. When you take a person's temperature—you are measuring. When a thermometer shows 41·5 °C (107 °F) it is pretty hot for atmospheric conditions, lukewarm for water in a bathtub, but when it relates to a person, that person is almost dead. Thus, when as a result of measurement we obtain figures, it should relate to a defined subject. In order to do so in this particular field, one figure, like body temperature, can neither reassure us nor should it alarm us.

Assessing the value of a maintenance service is comparable to appraising a supermarket which is providing all the services from A to Z, and at the same time trying to assess the effects it has on the surrounding community. It is not enough to show profits, the supermarket will also be assessed in terms of the community satisfaction that it can provide. In other words, showing a profit by operating the business efficiently is not yet conclusive. Do the customers get satisfactory service and does it allow *them* to work profitably? Or, are they waiting for service at times, and do they frequently have to spend extra time and effort in order to get the services from elsewhere?

All these are factors in any assessment of this particular supermarket. The analogy is quite close. A maintenance department has to manage its people, its facilities and its merchandise (spares) efficiently. It must be up to date in the selection it provides, it must allow quick service for those who want it (a "speed-line") and it must be able to cope with varied demands on stock items or specialised goods. It is set up to handle routine shopping lists but now and again a special request will appear.

Now if it were possible to visit all customers in their homes to follow up on the effect their shopping sprees have on their lives, we would be able to assess whether this supermarket is beneficial, neutral or detrimental to the community. This is the case for the average maintenance department. And these are the problems we are faced with when somebody asks "How's your maintenance doing?"

To arrive at a figure that would conclusively prove maintenance performance to be good or poor, we have to lay down rules for the game. What do we expect? Prompter service, more work done, less over-all costs? Some of these aims affect each other in an inverse relationship. If we want prompter service and faster response to calls we may need more people either at the central pool or in decentralised positions. Further measures would include localised sub-stores and better communication and transportation methods. This could improve service and could also raise the utilisation ratio of manpower, but it has to be financed. As a result, over-all operations may be less economical, *i.e.* the cost of an average maintenance hour may rise. It is also obvious that in economising on some measures the speed of service will be affected. For instance, in trying to reduce the stock of spares we could decrease inventory carrying

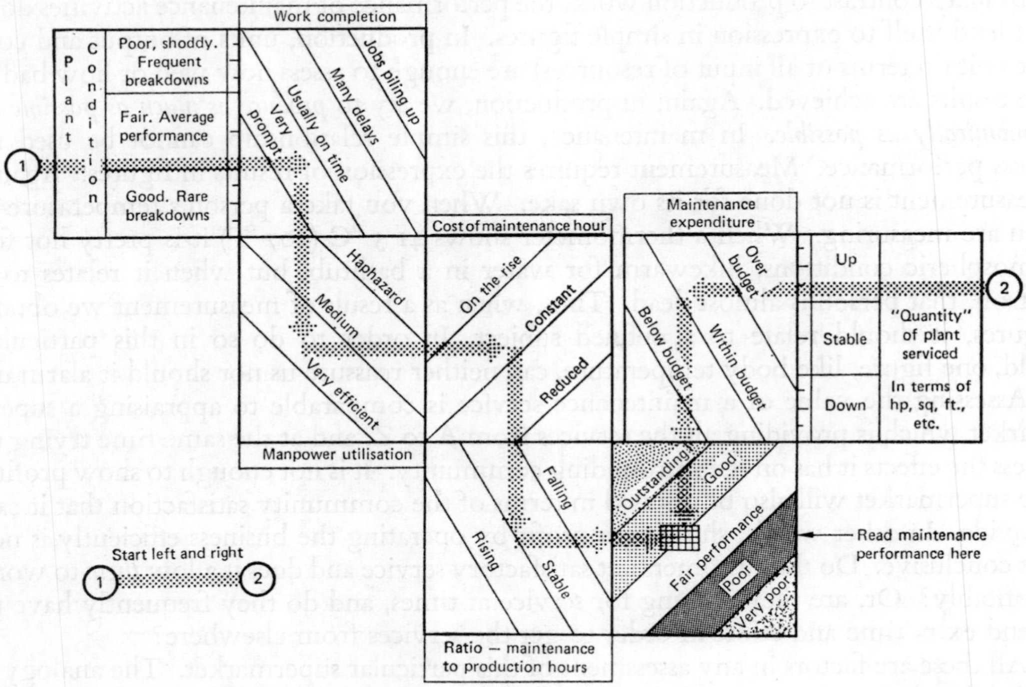

[*Courtesy McGraw-Hill Inc.*

Fig. 90. The "instant yardstick" multi-variable chart for assessment of maintenance performance.

costs. However, when occasional items are out of stock it may cause excessive delays, thus raising down-time costs.

Further examples are not necessary to show that appraisal cannot rest on any single feature. We could here adopt a sequential approach: if certain conditions are satisfied the final assessment will be favourable, but one unfavourable answer may spoil the over-all picture. To illustrate this approach a multi-variable chart (Fig. 90) has been developed by the author. This illustration is based on Phil Carroll's principles.[4]

This chart shows visually what we cannot grasp when surveying figures presented in a table. The interdependence of results also becomes obvious. This chart provides an opportunity for qualitative assessment, inasmuch as the squares can be entered to the appropriate spot representing the condition of the variable. Although the answer does not purport to be "scientifically" derived, it has its advantage in expediency.

"Performance measurement" *is expected* to indicate in two words both cause and effect *and* a comparison between past and present. Worst of all, it is expected to express in figures a constantly changing situation. Evidently, when we survey figures of the past and the present we expect to draw from it inferences for the future. Looking at the results we also try to understand their implication in light of probable future events.

One difficulty lies in the fact that people expect different answers, each person's expectancy reflecting his own particular background. In order to get a complete picture we have to distinguish between the following main features:

Group (*a*) Performance of the department.
Group (*b*) Assessment of the service.
Group (*c*) Expense justification.

These main groups can again be subdivided into their component factors as we shall see in Fig. 91.

The first aspect relates to the efficiency of the maintenance staff and the department. However efficiently the department is managed, it may still be working ineffectively on poorly planned schedules. Therefore, the second aspect, that of performance of service is to be appraised separately. In other words, efficient work and effective methods do not necessarily go together, but both must be present. When both aspects rate high, the expense justification follows automatically. This is the third aspect. It expresses in figures that a certain quality-level of maintenance management, providing a certain level of service, is costing a certain amount because of certain circumstances prevailing in this company.

The third aspect therefore ties the two previous ones to internal conditions which depend partly on external influences, *i.e.* deliberate management decisions. If, for instance, we mechanise certain operations that were previously done manually, there is more mechanical–electrical equipment to be maintained. When this process is carried further to an appreciable degree and the figures in the first two aspects remain the same, namely departmental efficiency remains high and the effects on equipment are still good, it indicates that maintenance is well able to cope with the added workload. This effect was expressed in 1963 in a simplified way by a maintenance manager of a British car-manufacturing company, "Since 1958 we have added two acres of manufacturing shops and our maintenance force has remained the same." How was this achieved? "By streamlining procedures and improving the supply of tools and spares."

Although excellent efforts have been made in recent years to combine overall-indicative aspects into one ratio,[5] such a factor cannot tell the whole story. We therefore proceed to investigate the three aspects mentioned above by detailing them further into workable ratios. An advantage of such an approach lies in the fact that we can directly observe the component factors and try to influence them by adopting corrective measures. Figure 91 shows how the three aspects are detailed into manageable ratios. The continuous use of all twenty ratios is neither recommended nor really feasible. Every company should select those that can most usefully be applied. Some

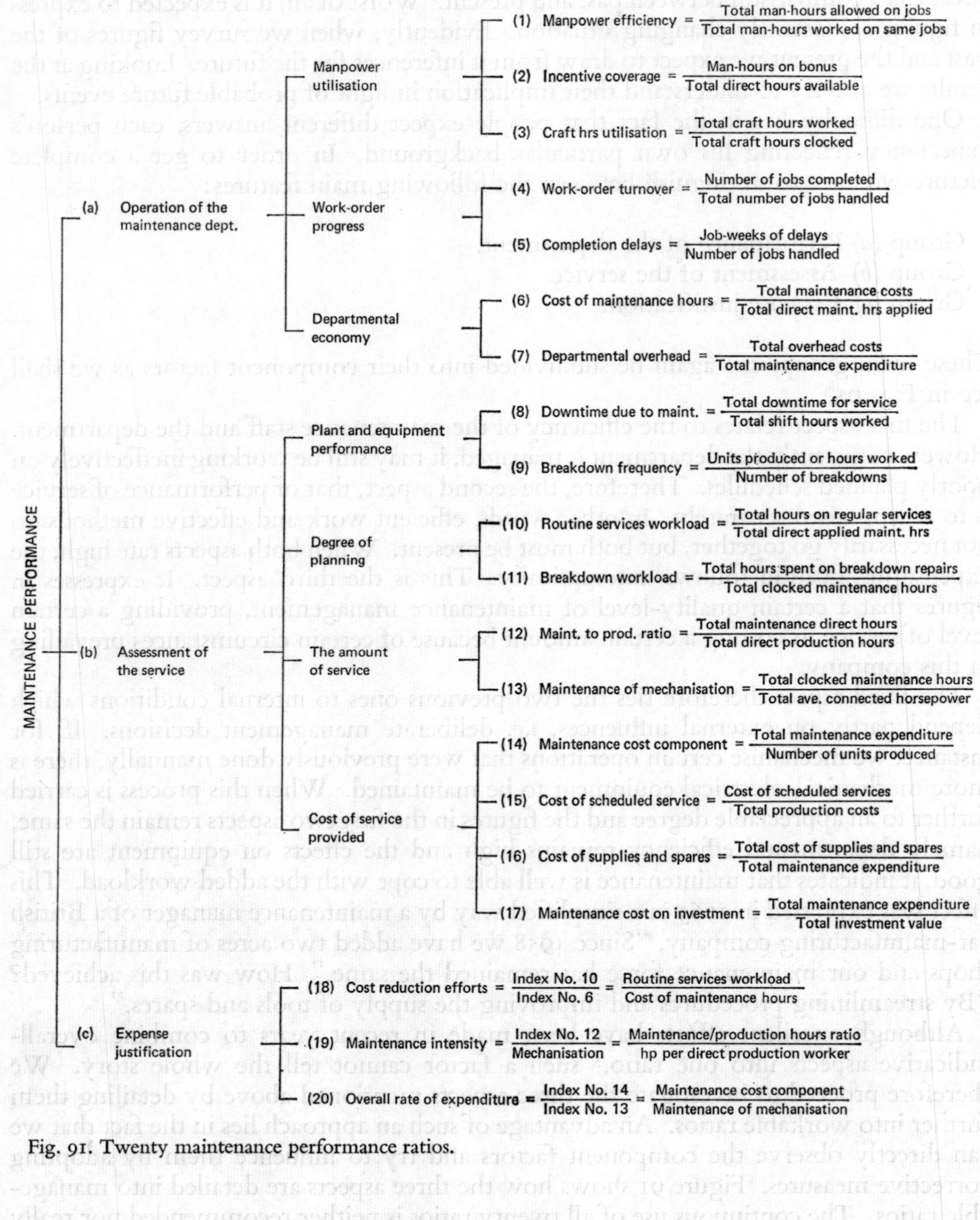

Fig. 91. Twenty maintenance performance ratios.

of the indices are only to be applied for limited periods, whereas others should be a permanent feature of reports. Let us discuss them one by one.

Group *a*: operation of the maintenance department

First and foremost we intend to discover whether the maintenance department, as a function, is operating efficiently. This over-all question is divided into:

(*a*) manpower utilisation;
(*b*) processing of work orders, and
(*c*) economy in operation.

The results obtained in this group will reflect the managerial ability of its chief.

Index 1—Manpower efficiency. Since efficiency is an input/output ratio, we assume the output in man-hours to be measurable as compared to the actual input, namely time spent on jobs. Thus manpower efficiency will be measured against a pre-set target. This ratio therefore presupposes a time allowance for all jobs, and cannot apply to tasks for which no standard time is set. Although this fact reduces its application, it can nevertheless serve as a useful indicator. Within certain qualifications one might overcome this disadvantage by post-estimating. However, the number of these cases should not exceed 10 per cent of total hours.

Manpower efficiency can sometimes be assessed from work-sampling studies exclusively based on effort ratings. This technique can serve as a good substitute for the above method. In that case the problem of pre- and post-estimates becomes redundant.

Index 2—Incentive coverage. This ratio is calculated on the assumption that the higher the percentage of hours worked on an incentive basis, the more work will be accomplished. There are two snags in this ratio, namely that the cost of attempting a high coverage may offset the benefits and that the coverage also reflects the ability of the estimator. In this way the performance of the department may hinge on his efforts.

A high percentage of coverage nevertheless indicates a certain level of competence available within the department. It is not uncommon to start with a coverage of 65 to 70 per cent and thereafter to obtain a slowly diminishing rise of 5 to 10 per cent per year. A 95 per cent coverage can be considered as an exceptionally high target.

Index 3—Utilisation of craft hours. All hours that can be directly ascribed to "productive" work are totalled here. Obviously, enforced waiting time, workbench cleaning, committee time, etc., will be excluded. Whatever the decision to include or exclude certain types of jobs, the percentage of hours accounted for out of the total clocked at the gate is a useful indicator. Time that we cannot account for is lost. Maintenance supervision must at all times endeavour to improve this ratio. The final figure will reflect upon the quality of supervision and the accuracy of the recording system.

Index 4—Work-order turnover. This ratio can only relate to work orders that represent individual requests occurring irregularly, such as repairs, improvements, transfer or installation of equipment, or overhauls. It cannot refer to lubrication, inspection and other recurring cyclic work on routine, planned or preventive maintenance. The denominator is made up of all orders that are "on-hand" at the start of the period. A widening gap between this figure and the jobs completed will indicate that a backlog of work is piling up. A rising number of incoming requests will not necessarily reflect badly upon the department's performance, but a low turnover will show that the team is unable to cope expeditiously with the load. A rising number of new jobs may however hint at the low effectiveness of the service. Whether this is so would be shown by Indexes 8 and 11. An alarming rise in the number of incoming orders would raise the suspicion that people are bothering maintenance with petty requests. A distribution chart along a severity scale (in hours) would show whether small jobs are too numerous.

Index 5—Completion delays. To refute complaints of long-overdue and neglected orders, it is sometimes advisable to group all incomplete (in-progress and pending) jobs into groups representing weeks since they were issued (*see also* Fig. 87). By multiplying their respective number by the weeks they have been outstanding, we get the total week-delays. By dividing this figure by the total of jobs handled (on-hand plus newly arrived), we get an average week-delay per order value. A value of less than one would indicate an average delay of under one week until completion. Values above two would justify the complaints. One way in which this could be affected by the supervisor is to handle all small jobs first. This in itself would not be an illusory advantage, since by logic small jobs should be fitted in between larger ones or as fill-in work. Such a procedure would also allow more time for the preparation of larger jobs.

Index 6—Cost of the maintenance hour. Some companies gauge their success or failure largely on this index. Cost estimates are also made on this basis as well as decisions to subcontract a job or to do it internally. Although there is a good reason to use this figure as a guide in cost estimating it may at times be misleading. The labour/materials/overheads composition of jobs differs widely and the staff may excel at some job. The combined effect of these three factors may cause some jobs to be either over- or underestimated. Time-consuming jobs like lubrication or fault-finding may be grossly overestimated while erection jobs will be underestimated.

A revealing improvement in the value of this index will occur as the department increases the amount of planned regular work. It therefore reveals the effects of improved management of the manpower, and the economic performance of activities.

Index 7—Departmental overhead. This ratio will not show sharp fluctuations. A certain percentage of overhead costs is unavoidable and not much can be done to reduce it. A great deal depends on what is included in the overhead, such as inventory carrying cost of spares, tooling, power and service facilities, clerical costs, etc. Every company must find its own level in this respect, since in trying to reduce overheads the planning function may be seriously affected. Whether reducing the ratio of overhead to direct expenditure is detrimental can easily be discovered when a large proportion of the cost of down-time is avoided by allowing inventory carrying costs to rise. In other words, perhaps a further stores overhead of £40 representing £2,000 of additional spares would reduce down-time by eight to ten hours. In practice, overhead costs should only be trimmed where they are not conducive to more effective work. It will be found that maintenance overheads are just as essential as direct maintenance costs are an overhead to production.

Group *b*: assessment of the service

This first group of seven indices focuses upon the way the function is operated. It preaches "get your house in order, first." The next group of ten indices attempts to assess the validity of the team's efforts. This is partly a qualitative assessment of the way work is carried out and partly an appraisal of the effectiveness on the equipment as evidenced by its performance.

Index 8—Down-time due to maintenance. The total amount of down-time is a result of unexpected stoppages, waiting for service, waiting for faults to be found and corrected and time required to start the machines after they have been serviced. It reflects the effects of thorough servicing or the lack of it, and also the promptness of response to calls. There is, of course, a great difference between a figure resulting from a large number of small waiting periods or a sum of two stoppages in which the spares had to be flown in from abroad. To reach any conclusion we would first have to examine both the composition of down-time and all features of the breakdowns. Nevertheless, as a value, we endeavour to minimise this figure and measures leading to this goal can soon show results. It is here that improved communications will show their impact. Better and faster diagnosis of the failure and the prompt availability of spares and tools, will prove their usefulness. Work study can markedly affect the total down-time, both when dealing with emergencies or in finding better ways to do routine jobs.

Index 9—Breakdown frequency. This ratio is intended mainly for units having a continuous throughput, such as pumps, mills, etc. Alternately, the work of a large group of identical machines can be checked by this figure, *e.g.* sewing-machines, groups of presses, automatic lathes and others. There is no benchmark for any of these, and here again the company has to find its own level. The larger the value of this index the more trouble-free running we have achieved. There is no objection to counting all mechanical units and multiplying them with shift hours, and using this figure as the numerator, however it is rather difficult to add into a single sum a number of sewing-machines, steam-calendars, pressing-irons and boilers. This figure is feasible for activities, such as steel mills, a fleet of trucks or a number of electric welders *i.e.* homogeneous groups.

Index 10—Routine services workload. How much time do we devote to regular scheduled activities? The assumption here would be that less than 25 per cent is too little and over 60 per cent is too much. By watching breakdown costs and down-time figures we could deduce empirically the required optimum percentage. This figure could include permanent assignments, such as air-conditioning and/or heating installations, the power-house, time spent by patrolling crews, in addition to the lubrication and inspection schedules.

The total of directly applied hours is the sum of hours spent on regular services, plus repairs, overhauls and plant expansion. These hours are all traceable and identifiable from job cards or service schedules. Some jobs done in the central shops, *e.g.* overhaul of dismantled parts, manufacture of components, will also fall in this category. However, unassigned time (waiting time) between jobs cannot be charged to any job. These will make up the difference between "applied hours" and "total clocked hours."

Index 11—Breakdown workload. It is significant to realise how much of the total hours paid for are spent on the repair of breakdowns. This will serve as a broad indicator showing a "penalty" figure that the company has to pay for some shortcomings. Identifying the causes (discussed in Chapter 8, pp. 171–5) will serve as a guide whether inadequate maintenance is the reason, or whether poor skills of operation, age and condition of plant or accidents are to be blamed. All jobs which appear unexpectedly and are deemed urgent, will fall into this category. Owing to the disruptive nature of such jobs they should be separated from regular jobs and the figures regularly scrutinised. Since the cost of breakdowns goes beyond the immediate cost of repairs and cannot very easily be identified, it would be extremely cumbersome to impose the recording of such costs on any system. Undoubtedly, an occasional illustration showing that the cost of replacing a £1·50 bearing was in fact £8·75, can serve as a valuable illustration. But an exercise of this sort cannot be pursued on a permanent basis.

Sometimes a factor can be calculated for multiplying the direct cost of emergency repairs, to give a total figure. If, for instance, replacing the £1·50 bearing took one man-hour, costing approximately £2 including shop overheads, and the total cost was calculated at £8·75 the factor would be $f(x) = 2{\cdot}5$. The extra cost would have arisen owing to machine and operator down-time and excess purchase costs. If such factors can be identified for certain jobs within a certain range from data obtained in several occurrences, it will probably apply in the majority of cases. This will allow management to realise the amount of losses that could be minimised by reducing breakdowns.

Index 12—Maintenance to production hours ratio. This ratio is used to express the amount of maintenance hours that are "spread over" production hours. It is as if we had a certain amount of butter to spread over slices of bread. The more slices we have, the thinner the layer of butter that is available to cover them. Within the same company a rising value can either indicate that we have more butter to spread, or that the equipment demanded more service. Once established, this ratio should either tend to decrease or remain stable. A variant to this ratio would be the number of hours per unit of product completed. In either case it is an indicator for the "*amount* of maintenance service" that we provide.

Index 13—Maintenance of mechanisation. The total amount of horsepower connected to the network represents the mechanical and electrical power employed on production and its auxiliary services. For a given "mix" of equipment, a certain age, its average condition and over extended periods, this index shows the rate at which the plant "consumes" maintenance services, the passage of time tends to cancel out the effects of improvements in applying the service owing to the inevitable deterioration of plant. Conversely, installation of new equipment tends to retard these effects although new machines are often more complicated and delicate than some old ones.

Index 14—Maintenance cost component. This ratio serves those companies which are sensitive to the cost of maintenance and often claim they can't afford a larger maintenance budget. When the maintenance component in a refrigerator is forty pence and the maintenance manager claims that he has to get by on a shoestring, a 25 per cent increase in the maintenance budget might increase the figure to fifty pence. On the other hand, it may never rise to this value, if, due to improved maintenance and less down-time, more units are produced.

Naturally, sometimes figures prove that maintenance costs are rising considerably when related to total production costs. Such may be the case in certain mining operations where two pence per ton may represent huge sums. In such instances a breakdown of maintenance costs must be made to discover the roots of the problem. Do spare parts and supply represent 50 or 60 per cent of total maintenance costs? If so, something must be wrong, possibly operators are not careful with equipment. Conversely, if manpower makes up the lion's share of total expenditure, it may indicate poor application of manpower.

Index 15—Cost of the scheduled services. This ratio should demonstrate the value of scheduled services. With a very modest rate of expenditure on regular services, a sharp reduction in breakdowns can be achieved. This is the cause of the misnomer, namely calling regular services "preventive maintenance."

When a certain saturation point has been reached, further expenditure on this type of service may not yield a reduction of breakdowns and may not lead to a decrease of other costs. This is the level that has to be watched. A certain amount of equipment can take only a limited amount of scheduled services. Services beyond that figure should be carefully controlled as it may represent over-maintenance. The question must be asked, what is done for the £25 spent per machine annually?

Index 16—Cost of supplies and spares. This figure should include inventory carrying costs and a total (where possible) of all issues from stores as well as the manpower taking care of the stores. Again, a total over-all figure that is suspiciously high should be investigated. The turnover rate of stocks can be applied to determine whether excessive inventory is on hand. There is no way to prescribe a figure for this ratio except to recommend that warehousing costs must be relevant to the value of stocks carried and that obsolete items should be "weeded out" during periodic surveys.

Index 17—Maintenance cost on investment. This serves as a broad indicator for every type of manufacture. A monthly expenditure of about 1 to 2 per cent will be quite common. This index may be used when certain periods are compared for budgeting purposes. Monthly figures may fluctuate widely and preferably quarterly or semi-annual figures should be calculated.

Group *c*: expense justification

Next we come to the group of indices that are both more sensitive to monthly variations and more comprehensive in their scope. By using four interrelated variables the over-all result is affected by each one independently. Such a computation results of necessity in an abstract figure that expresses several dimensions. Their meaning is entirely specific to a certain enterprise and to a point in time. Usefulness of the indices depends on a comparison between one period and the next. Since the characteristics of a company are "capsulised" in these double ratios, they can be used for inter-firm comparisons.

Index 18—Cost-reduction efforts. To what extent have efforts been made to produce an economical service? This ratio has somewhat the character of an output/input relationship. The output is expressed in the amount of scheduled workload, and the input in the hourly rate of the service. It assumes that over-all economy is achieved, up to a point, by reducing the cost of the maintenance hour. The value of this ratio will rise the more both of these objectives are being reached. Although providing the scheduled services should not be expensive in itself, its costs rise sharply if not properly applied or if a certain limit is exceeded. When this happens the cost of a maintenance hour will rise, lowering the value of this index.

Index 19—Maintenance intensity. The relationship between the consumption of maintenance services and the degree of plant mechanisation is here highlighted. This index measures the intensity of the service that is required to maintain a certain degree of mechanisation. With increasing mechanisation and a rising use of the average connected horsepower in operation, the final value will tend to decrease. This trend will be supported by the fact that the rise in applied maintenance hours will be much less steep in proportion to production hours, *i.e.* operating the equipment for a second shift will not require twice the normal amount of service hours.

For the purpose of comparison the degree of mechanisation expressed in terms of horsepower per worker is not applicable between different branches of industry. A shoe factory may be more highly mechanised than a furniture manufacture and this, at any rate, is meaningless. However, in a factory where shoes are mainly hand-made, the demand for maintenance will be smaller than in a mechanised one. A comparison between companies within the *same* branch of industry is therefore possible and indeed enlightening.

As in the case of other ratios, a comparison of values between periods is relevant.

This ratio will prove particularly useful for following-up on the gradual mechanisation and modernisation of a certain company. A fall in the over-all value will indicate a favourable trend. This may be upset by a progressive deterioration of the plant which comes about with increasing age and a replacement policy which is being implemented too slowly. Increased utilisation of plant (measured in production hours) will also raise the demand for maintenance services. These variables will affect the results. An improvement will be noticeable when machines start to be discarded and replaced, or by working overtime on existing machines.

Index 20—Over-all rate of expenditure. The objective in this ratio is to minimise the maintenance cost component of a machine-made product. Our success in achieving this objective will reflect our ability to manage this function economically. While pursuing a maximum degree of mechanisation we endeavour to keep the cost of maintenance from soaring. Therefore the value of this indicator should tend to fall.

It may be argued that the total kilowatt-hours consumed per worker during a period may also indicate the degree of mechanisation, or at least the consumption rate of power. Would it perhaps be advisable to use this figure in preference to connected horsepower?

Although there will be a tendency for the two variables to be closely related, the consumption of electric power is related to plant utilisation, *e.g.* shift hours worked. Instead of widening the scope of Index 19 it is preferable to relate two parameters that are linked from the start and thus increase the comparative response of this ratio. There may also be processes that consume a vast amount of power but are of a static nature, and suffering little wear and tear, *e.g.* furnaces or rectifiers. Thus we can much better associate mechanisation with an increasing mechanical power at the disposal of production workers, than with electro-chemical processes.

This wide range of indices is by no means exhaustive. Companies may develop their own ratios under particular circumstances and for specific demands. As can be seen, the significance of these ratios complement each other. None by itself can express the whole situation and the meaning of each must be viewed in the light of results of the others.

Again, it must be emphasised that occasional calculations are not sufficient. Certain selected factors must be regularly plotted on charts in order to discover trends. Often, more than one curve can be plotted on the same chart so as to show how the values vary along a time-scale, while the curves proceed alongside each other. This will indicate whether there is a relationship between them. In some instances a rising or falling trend in one variable may cause similar trends in other variables but with a time-lag, that is, a few months later. Such is actually the true purpose of these indices: to discover in what way measures taken in one area affect results in another, either favourably or unfavourably.

To avoid tedious and too frequent calculations, some figures can be added up quarterly or semi-annually and the averages assessed by dividing by three or six respectively. Watching for monthly fluctuations in some figures would be more confusing than useful. Then again, if a certain mechanism for obtaining these results is instituted, it might as well be left to run on its own in producing monthly figures. However, in interpreting them, either regular or moving averages should be worked out.

Knowledgeable authorities in this field have asserted that these indices should provide means to identify the guilty party among the host of variables.[41] They should at the same time allow prescriptive deductions to be made from their calculation. It can be argued that by failing to indicate specific remedial action, we have not accomplished much by following these indices. However, taking an analogy from the medical field, a doctor needs all the indications he can get on the condition of his patient in order to make an accurate diagnosis. If he then prescribes medicine and notices various improvements he will assume that his treatment proved successful. If the prescribed measures were not effective his medicine was faulty. Whether his diagnosis was a brain tumour instead of a liver ailment is irrelevant in the short term, as long as he has saved his patient. The vital symptom that had been temporarily overlooked will surely emerge in a subsequent routine test.

From watching the indices a maintenance manager should be able to draw conclusions and the symptoms will be adequate for him to prescribe remedial action. He will also, most probably, be familiar with his "patient" for a longer period than a doctor dealing with an emergency case. From the welter of background knowledge that cannot be encompassed by reports, and in spite of the obliteration of much detail, he will make the most of the indices to guide his decision-making processes.

Excessive overtime or rising cost of a maintenance hour does not and cannot indicate the solution: "Hire another foreman" or "Fire the storekeeper in the toolroom." The array of indices cannot do this since the figures are composite results of different factors. It is up to management to identify them and to explore by analytical means the reasons for the obtained results.

Decision-making involves an approach requiring both selection and quantitative assessment. By guiding this process into a defined, though admittedly wide, range of areas, the manager will possess a high degree of flexibility in his decision-making which allows him to select and combine his proposed action. The remedial action for different variables can thereby also be appropriately weighted.

If this procedure does not yet satisfy the exacting demands of some readers, we could enlist the services of the mighty computer. Since simulation and "modelling" are well accepted, if not very common techniques as yet, the day is not far off when these indices could be fed into a program that will reduce the interrelationships of these factors to single words or figures of assessment: 87, 6%, high, low, good, bad or indifferent. As long as we can specify to the computer what combination would be indicative of one state and what would indicate another, the computer can certainly do it with ease. However, in the last resort the making of a decision and its implementation still require human intervention. Or do we ever want to hold the computer responsible for "its" rash decisions?

Chapter 11

Reporting

GUIDELINES FOR REPORTING

Management's frequent lack of understanding of maintenance stems in no small measure from the fact that good maintenance reporting systems are hard to find. The daily rush of affairs in an average maintenance department is certainly not conducive to regular report-writing. Moreover, developing a good report format is difficult and ensuring a regular flow of data day by day, week after week, requires a dogged perseverance that is, to say the least, rather uncommon.

Whether poor communications between maintenance and management is a contributing *cause* for poor reporting or whether it is one of its *effects*, is a moot point. Whatever the answer, reporting has to be raised from its present dismal state. As long as the parties involved have not reached an agreement on the exchange of information, mutual understanding is impossible. From the maintenance manager's point of view it is a question of defining what he wishes to report and what he is able to compile. This is a major step in the right direction. From management's point of view the problem is the choice of certain information that will be useful and the cost involved in getting it.

We are here assuming that management is indeed anxious to get figures from maintenance. If this is not so, a maintenance manager would have first to satisfy his own needs in weekly or monthly returns and then would have to make the figures relevant enough to arouse management's interest when a report is submitted. It is invariably true that whenever hard facts are substituted for opinions and impressions, exchange of information becomes more valid, communications become important to both sides and misunderstandings tend to decrease.

It is essential to realise that reporting is a two-edged tool. It requires effort both at the preparation stage and also at the receiving end. Similarly, it is a commitment for both parties. Once you have started a reporting routine you have to maintain it. Once you are in receipt of a report you are assumed to have taken note of its contents, even if you disagree with its details or refute its claims and refuse the requests contained in it. Avoiding acknowledgment of a report that has been lying on your desk for some time provides no escape.

Owing to a combination of such factors information passes both ways sporadically and often under pressure of extreme urgency. The poorer the state of communication, the greater the need for good reporting. The chore of preparing reports is only eased by the realisation that it is in fact performing an important function. Unfortunately, there are too many managers who feel they can survive without summarising their activities even at quarterly intervals. It must, however, be very awkward when you are not able to account for your activities and those of your staff when asked to do so. It is therefore a definite advantage to prepare reports even if they are neither requested nor appreciated.

Having discussed the wholly negative side of the picture, let us assume that reports are favoured and that we are trying to perfect them. What are the ingredients of a good reporting system?

(*a*) An agreement on *what is needed* and *what can be provided* on a regular basis.
(*b*) A reliable and consistent *recording procedure* to obtain data.
(*c*) A summarising process and *staff* to do the job.
(*d*) A *simple format* for the report, containing concise and relevant information.
(*e*) A commitment *to read and to react promptly* upon receipt of a report.
(*f*) A good reason for maintaining a reporting system, *e.g.* the managing director's express wish.

Effective reports are usually brief. It is best to combine the presentation of figures with charts and appropriate remarks. The contents of a good report will be self-explanatory, avoiding the need to look up past figures or other reference material. The person submitting the report should preferably explain and interpret the results for the recipient and so allow the reader to focus his attention on the salient points. A good report will tend to demand action and the recipient will be left with no doubt in his mind regarding the steps that have to be taken.

A maintenance report will have to be all of this, only more so. Neither the writer nor the recipient have time to spare and both are, we hope, keen for action. If, on the other hand, the reader is left with an aftertaste of "so what?" the report is a failure.

Maintenance reports can be of two different kinds, namely regular periodic reports and special issues. The regular report should preferably be of a uniform format and length, with headings and blank tables pre-printed on one single sheet. If necessary the sheet can be a double-foolscap one folded in the middle, thus offering the use of two facing pages or the space of four pages in all. This size would be exceptionally long, and most commonly all the writing can be done on one typewritten page plus charts and their interpretation can be appended.

A recommended procedure is to use a carton folder as shown in Fig. 92 into which successive reports are filed. It should be of a distinctive colour such as orange or red. The facing page (the inside front cover) will hold the chart. The chart is a continuous one with a time-scale for a year or six months. On this chart the results for the month are plotted before the report is submitted. It contains past averages in order to avoid the need for looking up data elsewhere. The main report should be filed on the opposite page, stacked on top of the others that have preceded it. It is advisable to submit the folder with *a "remarks form" blank* inserted on top of the report itself, on which a number of headings are pre-printed, such as: "Remarks on operational/financial/performance figures," "Suggested lines of action," "Management's special request and point of view," "Budget appropriation sums and purposes" and "Notes for your information/action/discussion." The folder should be returned to the maintenance department by the middle of the month containing the filled-in "remarks form."

A frequent complaint of maintenance men is the fact that even though reports are submitted regularly or occasionally, there is often no reaction at all to their content. Feedback is essential. "If our reports are no good, why don't they tell us?" This, then,

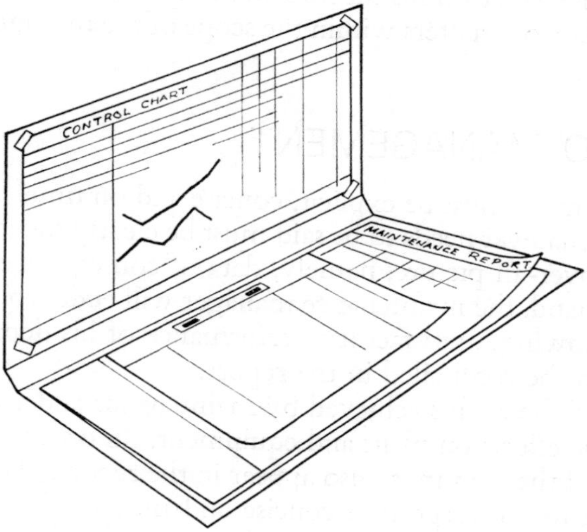

Fig. 92. A convenient folder for filing monthly reports.

is the reason for the above-recommended procedure. The folder would stay with the recipient not more than a week and returned with the relevant remarks. It can also be returned on the occasion of a brief personal discussion of its content.

A case can be cited where a rather highly paid consultant was hired to improve matters in the maintenance department. In order to ensure that the consultant earned his salary, the manager stipulated in his contract that he submit a monthly progress report. The first report was profusely acknowledged, the second earned a few comments, the third one got a note, and the subsequent ones were ignored. The consultant, having requested action that would enable him to proceed on a number of improvements and having obtained none, thereupon phrased his next report in totally negative terms, underlining in multicoloured pencils the problems and obstacles and interspersing the text with question marks and exclamation marks. The reaction to this report was swift. Although no names were mentioned in the report the manager requested further information on the quoted problems, the names of people concerned and initiated corrective action himself.

This example shows the value of a bold and imaginative approach to reporting. No effort should be spared to make a report convey its message even though the measures taken to achieve the objective may appear "corny" or unconventional.

In many ways reports should be cumulative, and should possess continuity. The value of a report is in relation to a point in time, where either the past or a future target is stated. A good report should read like a good radio or television serial, raising the question, "Well, what is happening this week (or month)? . . . what progress have we made?"

Distinction should be made between reports that are submitted by the maintenance manager to top management and reports that serve internally within the department. The latter may either serve to transmit information from one level to the next or to consolidate internal data on matters within the scope of the manager. This type will be discussed.

REPORTING TO MANAGEMENT

Reports to management must be explicit, concise and on time. There is no need for elaborate prose and whatever needs to be said must be clearly stated. Management will be interested in the over-all picture, namely, data of consequence to company operations. On the other hand, the maintenance manager will endeavour to present his case, where possible, as showing a satisfactory performance of his department. These two considerations have to be reconciled in the report.

As discussed earlier, the results achieved by maintenance and within the department must be related to the effects on plant and equipment. In order to convey a complete view, the other side of the coin must also appear in the report. The objective is for the relevant information to converge in a concise and meaningful way on one sheet of paper. Information that will complement maintenance data is as follows:

(*a*) *Cost figures* from accounting, relating to expenditure on payroll, spares and supplies, overhead, etc.
(*b*) *Production figures* on output, total production hours, down-time hours, breakdowns, scrap due to malfunctions and stoppages, etc.
(*c*) *Personnel data*, namely strength of workforce, recruitment and departures, vacancies, attendance and tardiness, absence for training, promotions, etc.

A compromise has to be decided upon between what is regularly and easily available, and what is desirable or absolutely necessary. It is indisputable that whatever the results in one respect their significance will be affected by results in another. We also want to avoid the situation where a production manager, for instance, may believe that he has good reason to complain. Certain small annoyances or one serious breakdown may appear important but, viewed in perspective, often do not justify the subsequent hue and cry.

It will immediately be apparent how wide the gap now is between what is ideal and what actually happens in practice. No one function can understand the other unless mutual exchange of information is taking place. The old parable about the elephant and the six blind men can here be quoted. As each of them touched a different part of the elephant they came to different conclusions as to its appearance. Clearly, it is indispensable to establish a procedure for collecting the necessary data and showing the complete picture. However ardently maintenance would wish to present a balanced view, this will depend on the co-operation obtained from accounting, production and personnel functions. This information is of importance to the maintenance manager as well as to top management. There may, however, be a difference in the way the two

MONTHLY REPORT
Maintenance Department

Date:
Serial no.

A. *Results in figures*

	Description	*Units*	*Monthly average*	*Data*	*Indices*	*(+) (–)*
a	Total departmental clocked hours	hrs				
b	Total directly applied hours	hrs				
c	Manpower utilisation ratio	%		a/b 100	☐	
d	Total direct + indirect payroll	£				
e	Cost of spares and supplies issued	£				
f	Total maintenance expenditure	£				
g	Annual expenditure date/target	%		g/target	☐	
h	Average cost of maintenance hour	£		f/b	☐	
i	*Reported downtime hours of all plant	hrs				
j	*Reported total cost of downtime	hrs				
k	*Reported total production units or hrs	hrs				
l	*Maintenance/production hours ratio	hrs		b/k	☐	

*Transmitted to maintenance by production and accounting departments.

B. *Interpretation* (variance from averages, target attainment, etc.)

I ______ Comments ______

II ______

III ______

C. *Workload situation* (completion of jobs and delays)

______

______

D. *Major points of interest* (difficulties encountered, action requested/pending)

a. People ______
b. Machines, plant, etc. ______
c. Tools and shop facilities ______
d. Spares and supplies ______
e. Budgetary ______
f. Communications ______

E. *Immediate future plans*

______

______

F. *Long-term planning* (required/in progress) regarding

______

Fig. 93. Format for a one-page monthly report.

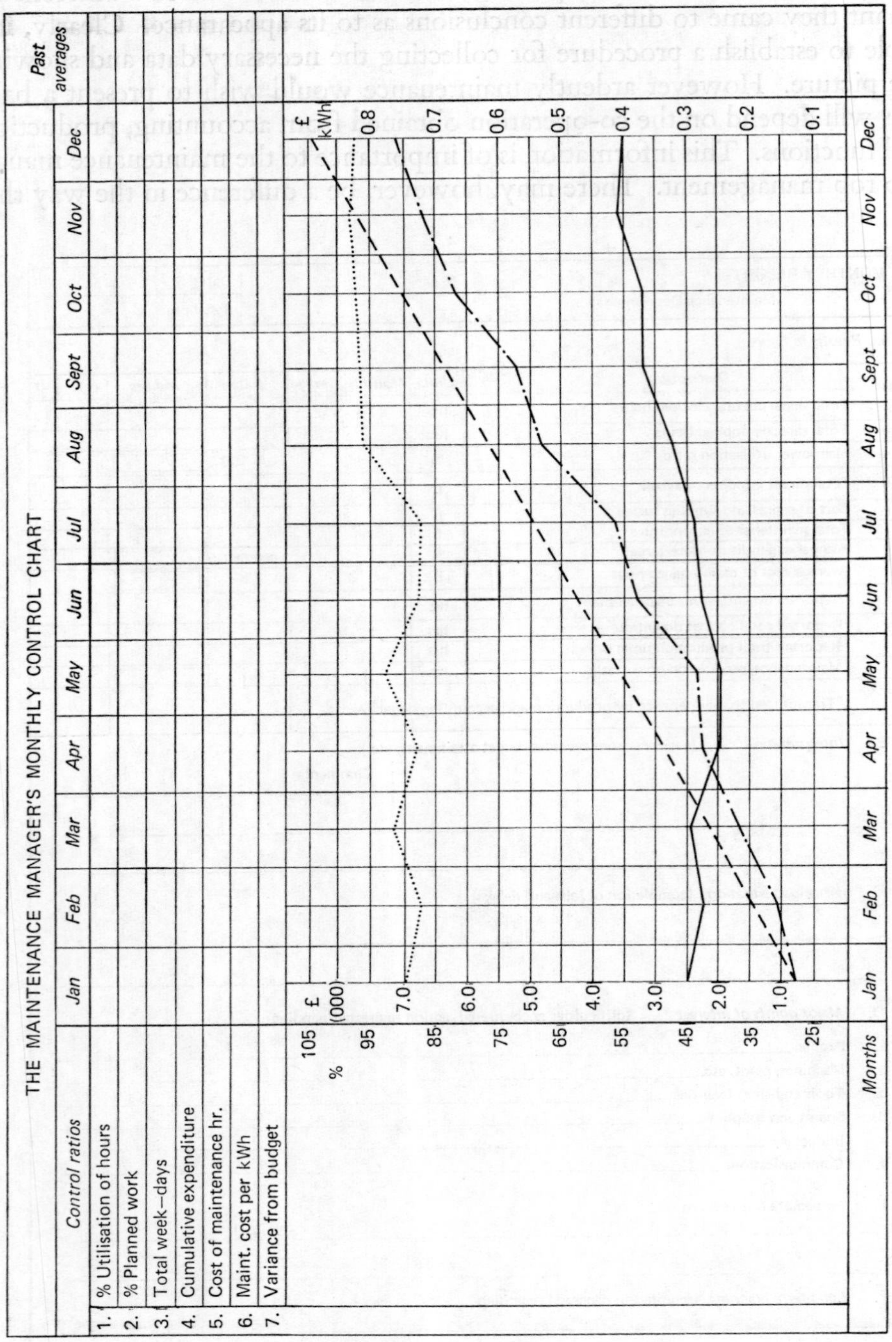

Fig. 94. A chart for the monthly posting of control ratios.

apply their judgment to the data obtained. It is advisable to channel the information through the maintenance manager, so as to allow him to better interpret his own operations. Significant relationships have to be discovered and transmitted to top management for necessary action.

Unless management is fully behind this endeavour, any attempt to implement a good reporting procedure is doomed to failure. Management must insist on compilation of data at all critical points and on a regular transmittal to maintenance. Understandably, time and money must be made available and not only grudgingly conceded.

If these guidelines are adhered to a clear picture will emerge about the activities of maintenance. To sum up, a view of the activities of the department is to be complemented by reports on plant operation, maintenance expenditure and personnel matters. A generally applicable monthly report form is presented in Fig. 93. The accompanying chart appears in Fig. 94.

INTERNAL REPORTING

Reporting to the head of a function means that he is to be informed about what is going on within his department. Many maintenance managers contend that they know exactly what they need to know. After all, they claim, they are constantly consulted on everything. Whenever they want further infoımation, there is Joe or Mac who can tell them. "Anyway, who can spend time writing reports? Getting the work done is more important."

People who can't be bothered about summarising their work "every now and then" will for ever be busy chasing information that should have been at the tips of their fingers. Who did what on which machine last year, and what is a reasonable estimate for job X? When did we last have that machine down for overhaul? How long did it take? And so it goes. But more unpleasant is the time when management wants to know what is happening to all those man-hours, to spares and supplies, and why machines keep breaking down. You're on the carpet and all the concentrated "instant-recall" information seems to evaporate. If this be the case, why should management have confidence in you? How can you justify, and how can you expect management to find justification for spending many thousands a year?

There is no alternative but to gather and compile data. The benefits of an internal reporting system will be felt at all levels. Foremen will realise the size of their operations, the factors affecting the results and will be able to assess what is important and what is not. In discussing their problems with other supervisors and the manager, they will be able to talk in concrete terms. Instead of planning in broad outlines, they will be able to marshal manpower resources, assign priorities and justify their decisions. Recording also imposes certain obligations, such as "clocking on" on jobs and "clocking off." Discipline in the department will thereby be tightened. Thus the benefit of reports is both in the process of getting the data and in its subsequent utilisation.

The mechanism of recording has been amply demonstrated in this text so as to allow us to turn our attention to the final presentation of data.

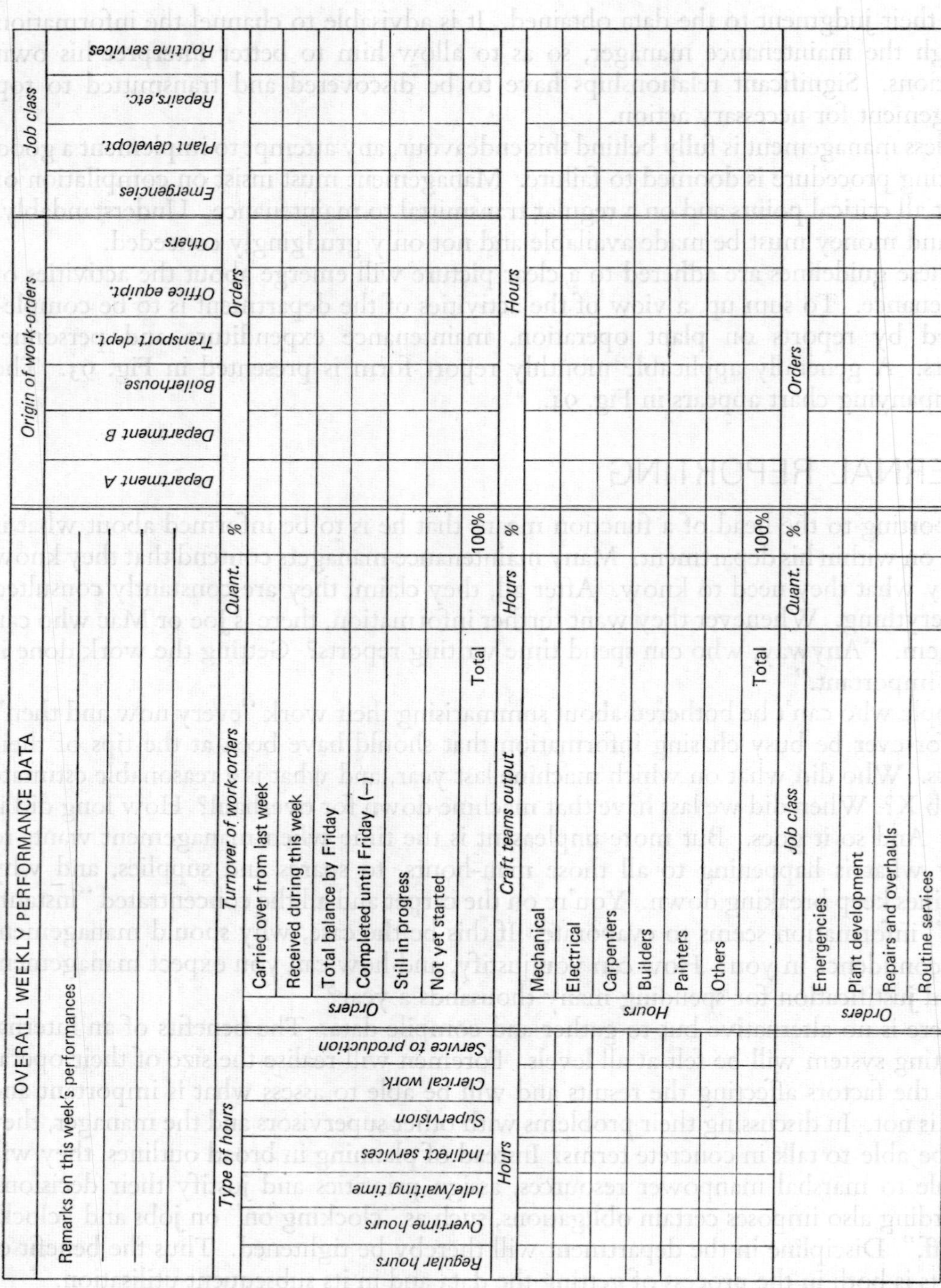

Fig. 95. Form for the multi-dimensional analysis of maintenance data.

Returning to our concept of multi-parameter presentation, we can assist in the interpretation of data by regrouping it into several classifications. The form appearing as Fig. 95 can easily be filled in by merely reading off hours from the foreman's daily report and checking on the work-orders situation. The advantage in this presentation is that it anticipates many of the possible questions; craft hours, for instance, are detailed in three different ways, namely by departments, by urgency and by type of hours. Similarly, if departments are charged for a certain total number of hours, it would be quite clear for what crafts and what class of services.

Figure 96 shows the procedure by which hours of several craft groups can be summarised daily and processed into a weekly hours analysis. Since the tabulation is both horizontal and vertical, it allows these sums to be posted week by week on charts. By providing this mechanism we get a visual picture of current trends. The advantage of this method lies in its continuity. Since figures are generated and recorded on the standard forms, it only requires the sums to be read off and posted regularly on the weekly hours analysis form.

These reports should be close at hand in the department. They should serve both as a reference and as a reminder. To facilitate a good over-all view a detailed management control board is here presented. Its use can be compared to a car's dashboard. Whatever critical figures are important to the driver will conveniently appear there. At the flick of a switch we can affect the way we proceed, light, music, ventilation, windscreen-wipers, etc. The same concept should serve us here as well. If we only had the same kind of positive-response controls!

The proposed management control board (Fig. 97) presents the critical data. Weekly or monthly progress will reflect the situation visually. There is a great advantage in this display compared to tabled figures. The significance of figures can be interpreted, magnitudes become meaningful and we don't have to read and interpret a series of five- or six-digit figures and compare them. Even after calculating and assessing their respective size, we have to consider whether there is reason to worry or to cheer. A further advantage in the use of charts lies in the fact that having found out the situation for our own benefit, we can communicate it to others with greater ease. The charts should be there progressing all the while. We will only refer to them when we need to do so.

Referring again to the progress of computerisation, it is feasible for this data to be fed into interactive systems and projected on to VDU screens upon demand. This type of "on-line" computer system could equally well show plotted curves of historical data on the screen as well as produce printouts through the appropriate terminals. Futuristic as it may sound it is already on our doorstep.

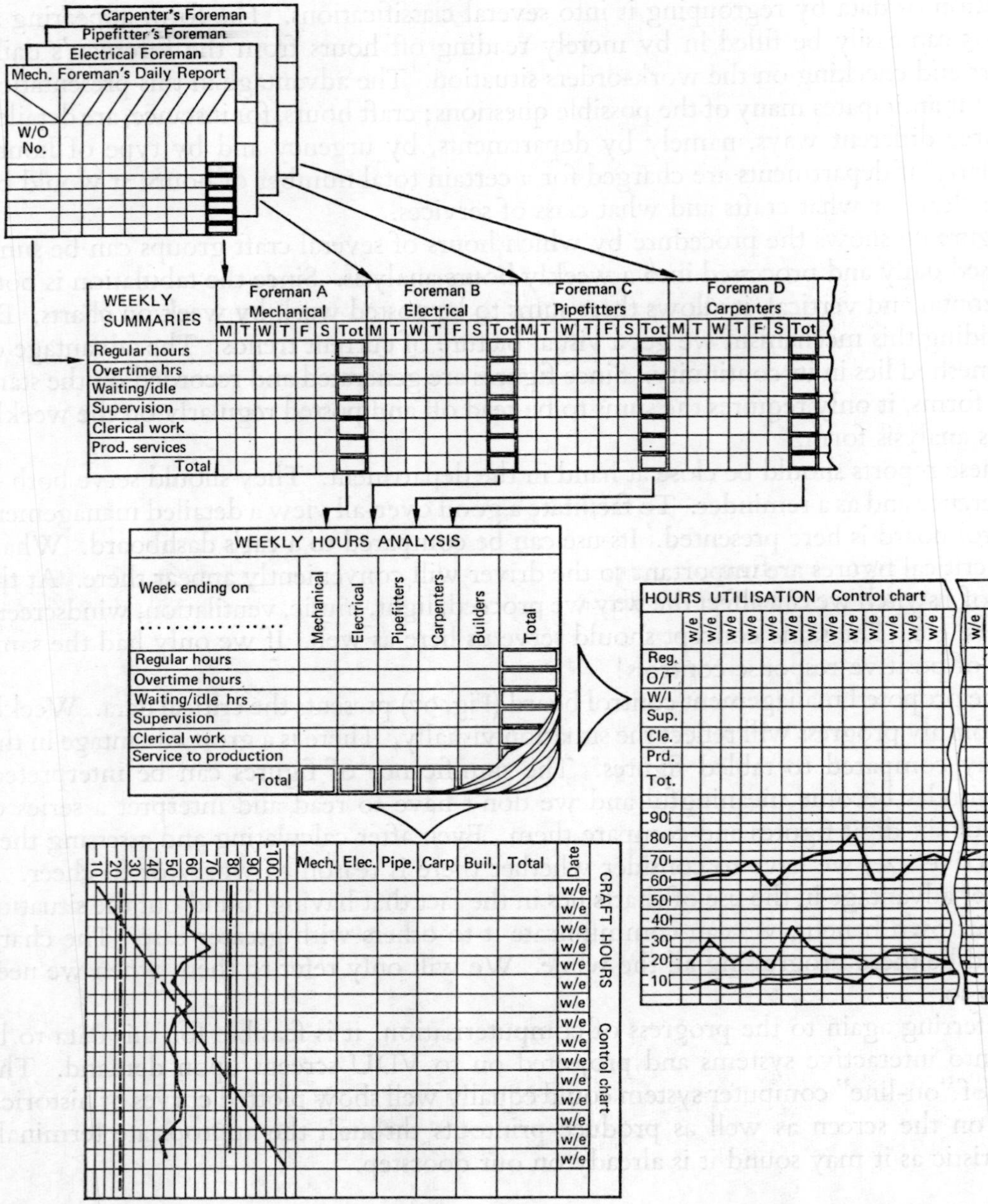

Fig. 96. The procedure for daily summarising hours of craft groups and processing into a weekly hours analysis can lead to the plotting of graphs in two directions.

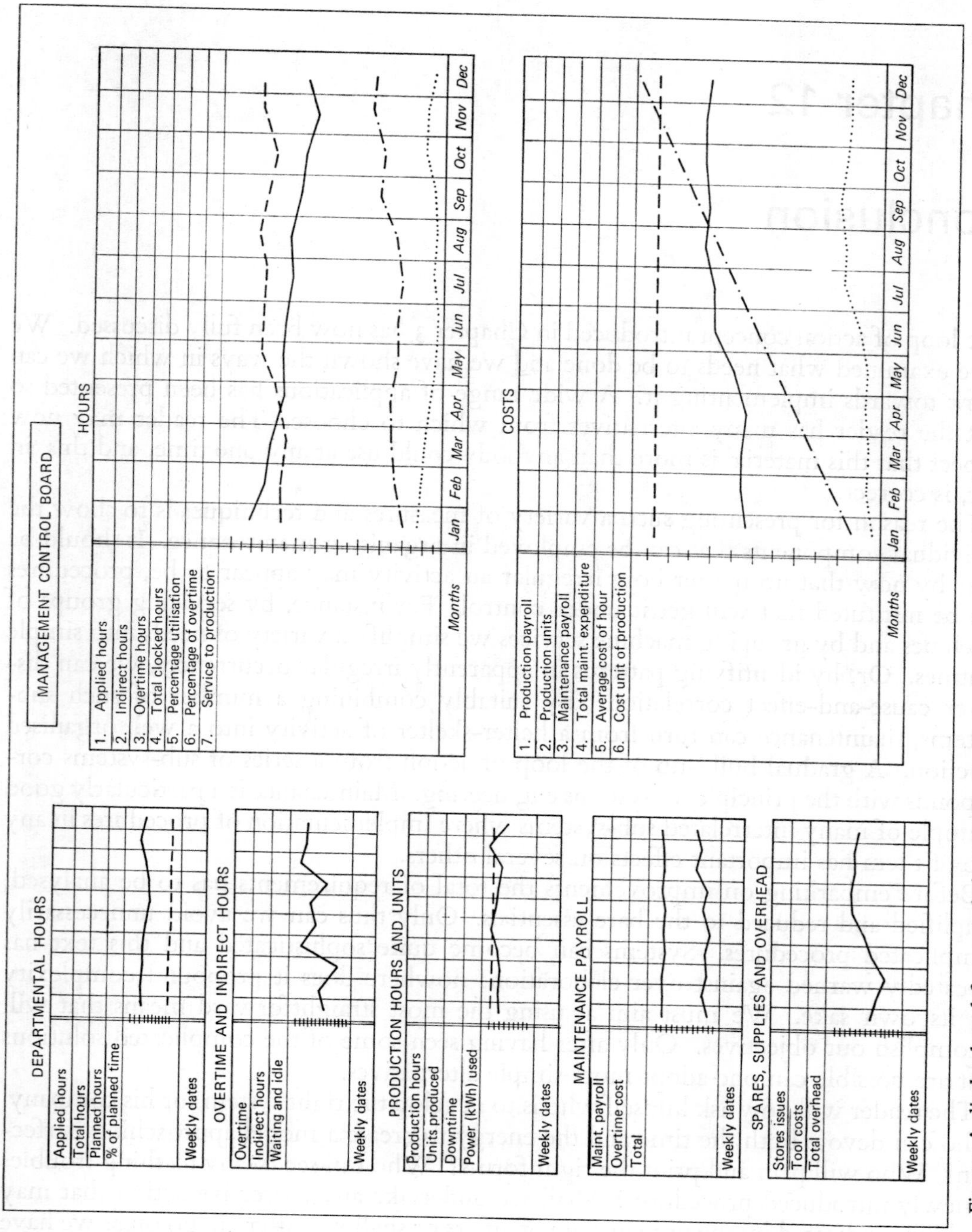

Fig. 97. A display board for control charts in the maintenance manager's office, showing weekly or monthly progress.

Chapter 12

Conclusion

The loop of action concept introduced in Chapter 3 has now been fully discussed. We have examined what needs to be done and we have shown the ways in which we can work towards implementing it. A wide range of applications has been presented so that the reader has many alternatives from which to choose. The reader may now suspect that this material is more than anybody could use at any one time, and this, in fact, is correct.

The reason for presenting such a variety of measures and techniques is to show the individual components that can be employed in organising maintenance. It should be clear by now that no matter how irregular an activity may appear to be, procedures can be instituted that will get it under control. For instance, by servicing groups of machines and by grouping machine services we simplify a variety of needs into simple routines. Or, by identifying patterns in apparently irregular occurrences, we can discover cause-and-effect correlation. By suitably combining a number of such sub-systems, maintenance can turn from a helter-skelter of activity into a well organised function. A gradual build-up of the loop of action from a series of sub-systems corresponds with the principles of systems engineering. Maintenance is a particularly good example of many interrelated sub-systems where implementation of procedures in any one of them has important effects on several others.

Before embarking on improvements the total of requirements has to be analysed, simplified and reduced to the bare essentials. Only thus can we avoid unnecessarily complicated procedures. Systems can become quite sophisticated and this text has repeatedly warned against over-elaboration; nowhere does it propound complexity for its own sake. We must aim at using the most straightforward means that will accomplish our objectives. Only after having seen some of the complicated solutions that are possible can one adopt more simple alternatives.

The reader will now ask himself who is to develop the right system for his company. Who can devote both the time and the energy to create a model approaching perfection? Who will plan and print the right forms? Who can see to any teething troubles in newly introduced procedures? Shall we undertake any corrective action that may be clearly indicated by the present state of affairs or shall we, after all, go on as we have done in the past—just patching up the more serious crises? Such worries and trepidations are quite understandable.

If you are a typical maintenance manager you are sure to be under constant pressure with hardly a moment to breathe. Yout next-in-line is probably your assistant manager or your best supervisor and they have their own share of the difficulties, and none of you can spare any time for procedural improvements. When you finally realise that nothing can be achieved without a corresponding input of effort, one of the following possible courses of action will be open to you:

(*a*) One of your department's capable foremen could be assigned to the task, supported by clerical assistance.
(*b*) A junior "systems assistant" from the white-collar staff of the company could be appointed to work under your direction. This would be very useful experience for an aspiring young executive.
(*c*) A consultant could be employed. This is an expensive solution unless the project is well defined in advance, but consultants could certainly be used to confirm or refute your own thoughts on what has to be undertaken.[52]

Whoever is entrusted with this task will have to work under your guidance and this will require a lot of your time and attention, but this is unavoidable.

A general movement towards better organised industrial functions is at present discernible in many industries. There is a greater availability of management know-how which has become more professional in nature. Pollution control and the escalating cost of industrial production have focused attention on the social responsibilities of industry. The phrases "quality of life" and "job enrichment" are now frequently used in public discussions. As a result we have become less tolerant of unsatisfactory working conditions and "meaningful work" is a prime concern.

However we find that the administration of maintenance is still sadly out of step with progress in other departments with which it works closely. We are surrounded by technical progress and we handle other people's technical development, but we have done relatively little to modernise our own field of work. The time is ripe for maintenance to employ some of the equipment that can reduce the daily drudgery and sheer leg-work associated with so many of our administrative functions. Office copiers, desk calculators, wall charts, accounting machines and electronic communication devices are but some of the products that could make our work more simple, more easy to control, more tidy and perhaps even more enjoyable. A more sophisticated means of administration will provide us with more satisfaction and it will allow us to contribute to the continued success of the enterprise. We will also be able to face the future with renewed confidence instead of risking "human obsolescence."

Since the mid-1950s much progress was made in raising the status of maintenance to become a recognised branch of management. We owe this progress to the development of other industrial management techniques and to the efforts of a number of dedicated pioneers who have worked for years to secure respect for the nature of our work. This has been achieved by means of technical publications and by organising conferences. Conferences on plant engineering and maintenance have been held annually in the U.S.A. since the early 'fifties under the guidance of the late L. C.

Morrow who also edited a comprehensive *Maintenance Engineering Handbook* in 1957.[9] Similar national and international conferences have also been held in the U.K. since 1961 under the auspices of Mercury House Business Publications Ltd.[16] Apart from a limited number of magazines which dealt with maintenance subjects, the bound volumes of the proceedings of these conferences were for some years the only source of up-to-date maintenance know-how.

More recently, maintenance has really achieved official recognition. As a result of a survey by a Working Party set up by the former Ministry of Technology the concept of *terotechnology* was brought into being. This term encompasses "a combination of management, financial, engineering and other practices applied to physical assets in pursuit of economic life-cycle costs; it is concerned with the specification and design for reliability and maintainability of plant, machinery, equipment, buildings and structures, with their installation, commissioning, maintenance, modification and replacement, and with feedback of information on design, performance and costs."[53] At present a Committee on Terotechnology advises the Department of Trade and Industry on matters relating to the new discipline and on ways to obtain its acceptance by industry. The British Standards Institute has also maintained for a number of years a committee on maintenance/terotechnology and has published a glossary of maintenance terms.[15] A British Council of Maintenance Associations (B.C.M.A.) has also been established and its honorary secretary, Dennis Parkes, M.B.E., is also the president of the European Federation of National Maintenance Societies (E.F.N.M.S.) which held its first congress in Wiesbaden, Germany, in 1972.

Courses and seminars on many aspects of maintenance have been offered for some years by leading consultancy firms and professional training organisations. These courses have now spread to several management training centres associated with universities and technical colleges. They are either residential or extra-mural and are sometimes organised under the auspices of the Institute of Mechanical Engineers or PERA (Production Engineering Research Association).

While the usual range of these courses includes organisation; planning, scheduling; cost reduction; spares controls; work measurement; fault diagnosis and replacement theory, some subjects are conspicuously absent, such as human factors and motivation; incentives (not necessarily time-based ones); management control information, and the application of computers to maintenance. These will probably be offered in the near future.

It is also regrettable that many of these courses are of the "come-on-along-and-listen-to-what-we-did-at-so-and-so" type with very little in-plant practice. Such practice would undoubtedly be more beneficial, though considerably more time-consuming, and such exercises would at the same time yield very useful case material that is all too often imported. The contention here is that maintenance is a subject that is best learnt by doing. Trainees should have the opportunity to see how theory is turned into practice instead of listening to lectures coming from behind lecterns.

We find it encouraging that an increasing number of colleges of technology are offering courses leading to H.N.C. and diploma qualifications in maintenance. Thus, finally, we can see an impressive upswing in activities on all sides. Recently there have even been suggestions at conferences that maintenance should really be considered as a partner of production. It appears, therefore, that our profession is at last gaining full acceptance and our specialised know-how is now available to all who wish to benefit from it.

Appendix I Cost analysis data for machine replacement

MACHINE TOOL REPLACEMENT ANALYSIS

	OLD EQUIPMENT	NEW EQUIPMENT
Name of machine	*Four engine lathes without bar equipment**	*Two turret lathes with hydraulic drive and bar equipment†*
Type, size, and horsepower	14in x 30in — 7½hp	Ram — 5·2½ — 20hp
Year manufactured	1939	1959
Year acquired	1939	1959
Original cost	£25,464	£54,000
Present book value	£ 0	£54,000
Present market value	£ 400	£54,000
Estimated life of usefulness		15 years

CURRENT COST COMPARISON (ONE YEAR)

Line number		*Avoidable costs or (savings)*
1.	Direct labour	£15,000
2.	Indirect labour	2,000
3.	Fringe benefits	4,500
4.	Spoilage in manufacture	1,000
5.	Maintenance — ordinary	100
6.	Maintenance — repair	400
7.	Power	(600)
8.	Perishable tools	
9.	Taxes and insurance	(1,600)
10.	Depreciation	(3,600)
11.	Other (list):	
12.	Total avoidable costs for one year	£17,200

* One operator — each machine

† One operator only — runs both machines

COMPUTATION OF AVOIDABLE COSTS INCURRED BY POSTPONING REPLACEMENT DECISION

	For one year	*For five years*(1)	*For ten years*(1)
Avoidable costs from current cost comparison			
line 12 opposite	£17,200	£17,200	£17,200
Avoidable costs from current cost comparison			
£17,200 times number of years	xxx	86,000	172,000
Factor for projected increase in labour and material cost			
£17,200 times 27.63%	xxx	4,752	xxx
£17,200 times 62.90%	xxx	xxx	10,818
Factor for projected increase in cost of new equipment			
£54,000 times 38.0%	xxx	20,520	xxx
£54,000 times 76.3%	xxx	xxx	41,202
Total avoidable costs incurred	£17,200	£128,472	£241,220
Proposed investment in new equipment	£54,000	£54,000	£54,000
Ratio of avoidable costs to cost of new equipment (for rating one replacement project against another)	31.9%	237.9%	446.7%

(1) Beyond first year

PROJECTION OF WAGE AND MATERIAL COST FACTORS

Year	*For year*	*Cumulative*
Current year	0	0
	%	%
1	5.00	5.00
2	5.25	10.25
3	5.51	15.76
4	5.79	21.55
5	6.08	27.63
6	6.38	34.01
7	6.70	40.71
8	7.04	47.75
9	7.39	55.14
10	7.76	62.90

Instructions: Apply the cumulative ratio (opposite the number of years, beyond the first year, for which a replacement decision would be postponed) to net of figures in Current Cost Comparison.

PROJECTION OF REPLACEMENT COST FACTORS

Year	*Actual replacement cost factor*	*Replacement cost factor — arithmetical average*
1	1.052	1.076
2	1.122	1.152
3	1.278	1.228
4	1.342	1.304
5	1.364	1.380
6	1.382	1.457
7	1.449	1.533
8	1.553	1.610
9	1.708	1.686
10	1.763	1.763

Instructions: Multiply today's cost for the new equipment by the replacement cost factor for the number of years, beyond the first year, that you would normally postpone the decision to replace.

Industrial World, December 1959

In this example the viability of replacement is projected for periods of five and ten years ahead for the proposed £54,000 machine (*see page opposite*). By using the compound interest scales of the two cost factors shown above, the avoidable costs have been calculated. Similar percentage ratios can then be obtained for other proposed investments to indicate the more favourable course to be adopted.

Project Number: 16 | Prepared by: J.J.A

Department: WAGES OFFICE | Date of preparation: 12·6·1973

Procedure/System/ Operation: PREPARATION OF PAYROLL

	Data	Existing Machine	Proposed Machine
(a)	Capital cost	£2,200	£3,000
(b)	Estimated life of machine	10 years	10 years
(c)	Estimated scrap value of machine at end of life	200	£ 500
(d)	Net capital cost of machine	£ 2,000	2,500
(e)	Annual depreciation (straight line basis)(*(d)/(b)*)	£ 200	£ 250
(f)	Expired life of machine	4 years	—
(g)	Book value of machine (*(a) less (f)x(e)*)	£1,400	—
(h)	Present market disposal value of machine	£ 500	—
(i)	Market value estimate for one year hence (new machine not being considered for disposal)	£ 200	—
	Annual Direct Costs	£	£
(j)	Operating labour:	600	600
(k)	General supplies	100	100
(l)	Equipment costs:		
	(i) Depreciation (*(h)* less *(i)*)	300	—
	(ii) Depreciation percent (proposed machine normal basis)	—	250
	(iii) Interest on capital (@10% p.a.)	*50	**250
	(iv) Maintenance	300	75
	(v) Electrical power	50	50
(m)	Space costs	50	50
(n)	Loss on disposal of existing machine (written off against proposed machine over the remaining years of life of the existing machine)	—	150
(o)	Total costs	1,450	1,525

This is another example showing the factors involved in the calculation of direct costs of ownership of an existing and a proposed machine. These costs are then considered in a further comparison of output benefits to be derived.

Appendix II

Technical and cost factors in a machine replacement check

A. TECHNICAL FACTORS

1. Is the present equipment worn out?
2. Is it obsolete?
3. Is it inadequate from the standpoint of:

   (*a*) Range of size of work.
   (*b*) Speed of operation.
   (*c*) Accuracy of work.
   (*d*) Strength or rigidity for heavier operations.
   (*e*) Rate of output.
   (*f*) Insufficient power?

4. Has it been made unsuitable by other changes in equipment in the plant as for example, the setting up of a product line of manufacture, or the purchase of other machines working to closer tolerances?
5. Can its operations be more readily done if combined with other operations on an automatic machine?
6. Does it have the controls, special attachments and safety features of new types of equipment?
7. Will a new machine not only do the present work but also other kinds of work which the present machine cannot handle?
8. Will a new machine replace hand operations or bench work?
9. Will a new machine have special advantages from the standpoint of:

   (*a*) Ease and speed of set-up.
   (*b*) Convenience of operation.
   (*c*) Safety, such as guards, stop buttons, etc.
   (*d*) Reliability in performance?

B. COST FACTORS

1. Is the cost of keeping present equipment in good repair too high?
2. Would the cost of altering it for new work be too great?
3. Will spoiled work be reduced by the greater accuracy of new equipment?
4. Will a faster rate of production be obtained?
5. Will one new machine do the work of two or more of the existing machines?
6. Can machine operatives be substituted for skilled craftsmen, thus lowering labour costs by the change?
7. Will one operator be able to tend two or more of the new machines?
8. Will maintenance cost of the new equipment be less than that of the old one?
9. Will the new machine save manufacturing space?
10. Will it be conducive to higher worker productivity?
11. Will it provide the basis for better service to the customer?
12. Will the product for which the machine is to be procured continue to be made for a considerable time?
13. Should that product be later dropped, will the machine fit into other work?
14. How soon must the machine pay for itself to justify its purchase, especially if products may change?
15. How many years of effective service may be expected from the machine?
16. How will the cost of operating the new equipment be charged to the product?
17. Are funds available for the purchase of the equipment or can the investment be specially financed?

Appendix III Development of bench-mark job times

Task Area: General Installation

Code: 0795
Craft: Electrical

Group D (0.5) 0.7 (0.9)	Group E (0.9) 1.2 (1.5)	Group F (1.5) 2.0 (2.5)	Group G (2.5) 3.0 (3.5)
0790-6 – medium-size junction box, 4 tapped holes, 26 wires No.12, screw clamp connection, *mount* and *connect* 0790-7 – medium-size junction box, 4 tapped holes, 17 wires No.12, screw clamp connections, *mount* and *connect.* 0790-10 – wires, 14-No.12, 15ft, *measure,* cut off, identify, and *install* in 15ft of conduit	0790-16 – conduit, 15ft-1¼in, 2-30° bends, 2 condulets, 2 nipples between junction boxes, *prepare* conduit and *install:* 2 men 0790-2 – medium-size junction box, 4 holes, 37 wires No.12, crimped connections, *mount* and *connect* 0790-5 – medium-size junction box, 4 holes, 28 wires No.12, crimped connections, *mount* and *connect* 0790-9 – wires 22-No.12, *measure,* cut, identify, *install* in 15ft conduit	0790-15 – conduit, 35ft-2in, 2-30° bends, 2 condulets, 2 nipples between junction boxes, *prepare* conduit and *install:* 2 men 0790-17 – conduit, 15ft-1½in, 2-30° bends, 2 condulets, 2 nipples between junction boxes, *prepare* conduit and *install:* 2 men 0790-8 – wires, 37-No.12, *measure,* cut, identify, *install* in 35ft conduit	0790-3 – medium-size junction box 4 holes, 85 wires No.12, crimped connections, *mount* and *connect* 0790-11 – wires, 54-No.12, *measure,* cut, identify, *install* in 80ft then 50ft conduit 0790-19 – medium-size junction box, splice, No.12, wire, *make 54*

Task Area: General Installation

BLOCK 1 BENCH MARK ANALYSIS SHEET CODE 0790

Description: Medium-size junction box, 4 holes, 85 wires No.12, crimped connections, mount and connect | *Date: 4/26/61* | *B.M.: 0790-3*
Craft: Elect.; Gen. Install.
Dwgs: None
No. of men: 1 | *Analyst: W.M.* | *Sh. 1 of 1*

Line	*Men*	*Operation Description*	*Reference symbol*	*Unit time*	*Freq.*	*Total time*
1		Mount medium-size box	750.0207			0.3243
2		Select proper wire	13.0002	0.0035	85	0.2975
3		Move marker on wire	720.0660	0.0094	85	0.7990
4		Cut off 85-No.12 wires	720.0101	0.0021	85	0.1785
5		Skin 85-No.12 wires	720.0211	0.0023	85	0.1955
6		Connect 85-No.12 wires	720.0323	0.0110	85	0.9350
7						
30						
Notes:			Bench mark time			2.7298
			Standard work group			G

BLOCK 2 UNIVERSAL STANDARD DATA CODE 0720.02

SKINNING

Symbol	*Skin electrical conductor No.18 through MCM*				*Hours*
720.0211	Single conductor cable	Small	No. 10, 12 gauge and smaller		0.0023
720.0212		Medium	No. 4, 6, 8 gauge		0.0034
720.0213		Large	No. 2 through MCM gauge		0.0076
720.0214	Multiple conductor cable	2	No. 10, 12 gauge and smaller (Romex)		0.0031
720.0215		Conductor	No. 10, 12 gauge and smaller (BX)		0.0077
720.0216	Does not include inside conductor	3 Conductor	No. 4, 6, 8 gauge	Unarmoured	0.0086
720.0217				Armoured	0.0167
720.0218			No. 2 through MCM gauge (4-6ft) armoured		0.0558

Table values are for skinning only. They include tool and material handling, but no cutoff.

BLOCK 3 OPERATION SYNTHESIS CODE 0720.02

Symbol	*Ref.*	*Operation or element description*	*TMU*	*Freq.*	*Total*
720.0211		Skin wire No. 10, 12, and smaller (levelled hrs. 0.0023)			
	05.0004	Handle knife	100.5		
	04.0001	Handle wire	34.2		
	A	Skin end of wire	96.5		
			231.2		

BLOCK 4 ELEMENT ANALYSIS CHART CODE 0720.02

Description – left hand	*No.*	*LH*	*TMU*	*RH*	*No.*	*Description–right hand*
A. Skin end of wire with knife – No. 10 or smaller						
Move to area		M10B	18.7	M16C		Move knife to work
			16.2	P2SE		Align
			2.9	M1B		
			16.2	AP1		
			5.4	T90S		Skin wire
			16.2	AP1		
			7.5	D2E		
			13.4	M12B		Move knife away
			96.5			

A sample page from a UMS manual. [*Courtesy: H. B. Maynard & Co. Ltd.*

Appendix IV

A review of personnel policies, practices, working conditions and morale-boosters

1. MANAGEMENT POLICIES

(*a*) To what degree has management defined its attitude to maintenance? Has management laid down policies concerning production plans *versus* maintenance service? To what extent have these policies been brought to the attention of the maintenance personnel?
(*b*) Does occasional conflict occur between priority of production and maintenance?
(*c*) How does one get an extra budget allocation for urgent maintenance jobs? Or is it a hopeless task?
(*d*) Is knowledge of the maintenance programme a clearly defined part of a production supervisor's job?
(*e*) Are there clearly defined policies concerning selection of personnel for the maintenance workforce? How were your present maintenance foremen promoted to their jobs? How many years ago?
(*f*) Is training and keeping up to date one of the policies regarding maintenance personnel? How is this accomplished?
(*g*) Is there a clearly defined policy concerning upkeep and planned obsolescence of plant? Are they well known to maintenance?
(*h*) Are there directives as to permissible alternatives concerning methods of installation, repair and shutdown or replacement for most plant units?
(*i*) Has a priority guide for critical plant been established for cases of simultaneous breakdowns?
(*j*) Is there a well-known procedure for spare-parts consumption as well as reconditioning of dismantled parts?
(*k*) Is a budgetary control exercised? Is it a kind of post-mortem by the accounting function or does it attract the alert interest of management?
(*l*) Is there occasional interest shown by management concerning over-all performance of maintenance and its problems? Or does the statement apply: "Management doesn't care a bit what happens to us in maintenance?"

2. PERSONNEL MANAGEMENT PRACTICES

(*a*) Do you consider selection, placement and promotion practices satisfactory? Are they formalised and adhered to and well known to workers?

(*b*) Have these practices been established in consultation with senior maintenance supervisors?
(*c*) Is there any check on how well these personnel procedures operate? (Opinion surveys, sample checks, etc.) Are they casual or have they become more formalised?
(*d*) Is there a formal procedure for follow-up of newly employed personnel?
(*e*) Is there an incentive scheme in operation? What is its basis? Are maintenance foremen included in the scheme? (Allowed time, increased output, plant utilisation or a "blanket" bonus.)
(*f*) For how long has it been in operation? What changes have there been made lately, if any and on whose initiative?
(*g*) Is there a general satisfaction with the scheme? How do you know?
(*h*) Do you consider the remuneration side well taken care of?
(*i*) Have earnings increased lately and by how much? Has it been in line with that of production workers and has it kept pace with efficiency of the plant? Any complaints in this respect?
(*j*) Specify semi-financial incentives: subsidising outside training *yes/no*, extra leave following overtime and special jobs ——, means of transportation during overtime ——, extra sets of clothing and safety apparel ——?, etc.
(*k*) Are there certain measures for non-financial incentives applied? Titles ——, badges ——, authority ——, representative posts ——, special appointments?
(*l*) Does this statement apply: "We have tried incentives in the past but results have not been encouraging so we have given up trying. Anyhow the men don't go for it?"
(*m*) Is supervision fairly smooth? Are there cases of complaints about foremen?
(*n*) What form does supervision take? Giving orders and instructions before work starts ——, inspecting finished work ——, general round of inspection ——, getting reports of completed work ——, reprimands when something goes wrong ——? Or is it barely noticeable?
(*o*) Would you define supervision as formal, strict, friendly or advisory in character?
(*p*) What functions do foremen have to carry out? Clerical ——, Supervisory ——, Stores ——? Does this leave enough time for foremanship?
(*q*) Does a foreman frequently have to call upon his formal authority to get something done?
(*r*) What are the figures for lateness ——, absenteeism ——, turnover ——? Are they regularly completed ——?
(*s*) Do workers willingly take nightshifts? or week-end work ——?
(*t*) Any cases of disciplinary action required lately?
(*u*) Is there a "Book of Regulations" available for maintenance personnel where policies and procedures are outlined, as well as special terms regarding maintenance workers?

3. WORKING CONDITIONS

(*a*) Apart from difficult conditions in some production processes, would you say that maintenance work is carried out in satisfactory physical surroundings?
(*b*) Are maintenance shops well lit, clean, fit for work, with paved floors? Are workers provided with adequate, well-kept and up-to-date tools?
(*c*) Are tool and spares stores well equipped and managed, or do they cause irritation because of slow and poor services?

(*d*) Are rest periods well organised and are there any services provided during breaks? Are there canteen facilities available on late shifts or on week-ends?
(*e*) What are the measures taken to ensure safety at work, both for maintenance workers and operating personnel? Posters ——, electrical and pipe markings ——, colour schemes ——, any other measures?
(*f*) What safety apparel is provided?
(*g*) Is safety part of a foreman's job?
(*h*) Have safety programmes been regularly undertaken?
(*i*) What is the accident rate —— and severity —— for maintenance workers? How does it compare with the rest of the plant?
(*j*) How would you rate housekeeping within the maintenance shop areas including benches, chests and lockers?
(*k*) Do maintenance workers clean up work locations after completion of a job? Have there been any complaints on this issue?
(*l*) If you consider tidiness and orderliness a measure of morale, how would you rate your maintenance workforce?
(*m*) Are there any welfare and social activities going on within the maintenance group or in participation with production personnel? If so, is there genuine interest shown?

4. RELATIONSHIP AND COMMUNICATIONS

(*a*) Is there a feeling of management leadership in the field of all-round harmonious relations? How is it implemented and evidenced?
(*b*) Is the organisation clearly defined by procedures, and is there an occasional exchange of ideas between functions on problems of authority and organisational procedures?
(*c*) What level does the department head report to? To what effect, if any? How many levels are there in the maintenance department itself? Do they report in any way, or only if something goes wrong?
(*d*) Is permanent and temporary responsibility fairly well distributed, so as to lead to a feeling of general participation? Or is most of the team in an ill-defined pool of functions?
(*e*) Are there some "sheltered" jobs, carrying titles or other benefits which are coveted?
(*f*) Can relations between maintenance and production be described as smooth? Any exceptions, whether persistent or occasional? On whose side?
(*g*) What is the relationship between maintenance and the personnel department, transportation, quality control, production planning, accounting? Noticeable in any way or non-existent?
(*h*) How do all functions regard maintenance? With respect ——, indifference ——, ridicule ——. How do you know?
(*i*) What means if any are there to ensure two-way communications? Through supervisors and/or foremen?
(*j*) Are there any difficulties of authority directly connected with work arising on the shop floor? Any conflict between production and maintenance priorities?
(*k*) Is there much inter-level communications within the maintenance functions? Is it on a regular basis?
(*l*) Are there proper channels for airing opinions? Do you consider free discussions beneficial? Would an increase in communications make it a nuisance?

(*m*) Is anybody hampering good relations? Are workers encouraged to pass on and discuss technical information?
(*n*) What are relations with union representatives?
(*o*) Are shop stewards co-operative? Do you consider them as advisers, meddlers, obstructors? On any of the recent discussions has any effort been spent to make them see management's point of view?
(*p*) Is the shop steward kept constantly busy by workers referring to him?
(*q*) Are frequent cases of complaints and grievances filed with the shop steward?
(*r*) How many of the maintenance supervisors, foremen and engineers have had any formal human-relations training?

5. MORALE-BOOSTERS

(*a*) What techniques are deliberately undertaken as morale boosters?
(*b*) Is there a works bulletin or any other house organ? Does it relate mainly to new personnel and social activities or does it mainly deal with industrial and technical matters?
(*c*) As an alternative, are notice-boards utilised to convey information?
(*d*) Is any publicity given to maintenance activities?
(*e*) Are maintenance workers given any plant-wide recognition for outstanding jobs?
(*f*) Is there a suggestions scheme in operation? Is it plant-wide and how well do maintenance workers participate?
(*g*) Is there an attempt to raise efficiency by means of the suggestions scheme or is it merely a bother?
(*h*) What are the number of suggestions from maintenance, how many are accepted and implemented and what is the yearly sum paid out in prizes?
(*i*) If there has been a failure in this scheme have the reasons been analysed?
(*j*) Is there an apprentice scheme in operation? How well does it operate?
(*k*) Is there in-plant training for maintenance workers? Has an attempt been made to measure its effects?
(*l*) What is the relation between training, seniority, upgrading and promotion? Coincidental or deliberate?
(*m*) Would recreational facilities be welcome?
(*n*) Have any requests of a general nature been made recently? What was the action taken?

Bibliography

BOOKS

1. Alford, L. P. and Bangs, J. R., *Production Handbook*, Ronald Press, New York, 1955.
2. Blanchard, B. S. and Lowery, E. E., *Maintainability, Principles and Practices*, McGraw-Hill Inc., New York, 1969.
3. Caplen, R., *A practical approach to reliability*, Business Books Ltd., London, 1972.
4. Carroll, P., *How to chart time-study data*, McGraw-Hill Inc., New York.
5. Corder, G. G., *Maintenance—techniques and outlook*, British Productivity Council, London, 3rd ed., 1968.
6. Jardine, A. K. S., editor, *Operational Research in Maintenance*, Manchester University Press, 1970.
7. Lewis, B. T. and Pearson, W. W., *Maintenance Management*, John F. Rider Publications, New York, 1963.
8. Miller, E. J. and Blood, J. W., editors, *Modern Maintenance Management*, American Management Association, New York, 1963.
9. Morrow, L. C., *Maintenance Engineering Handbook*, McGraw-Hill Inc., New York, 1957.
10. Newbrough, E. T., *Effective Maintenance Management*, McGraw-Hill Inc., New York, 1967.
11. Sward, K., *Machine tool maintenance*, Business Publications Ltd., London, 1966.
12. *Control of Maintenance Costs*, Research Report No. 41, American Management Association, New York, May 1964.
13. *Annual Plant Engineering and Maintenance, Proceedings and Papers*, arranged by Clapp and Poliak, 245 Park Avenue, New York, 10017, 1952, *et seq.*
14. *MAPI Replacement Manual*, Machinery and Allied Products Institute, Chicago, Illinois, 1950.
15. *Glossary of terms used in maintenance organization*, British Standard 3811: 1964 (under revision at time of going to press).
16. Clements, R. and Parkes, D., editors, *Manual of Maintenance Engineering Combined*, Mercury House Business Publications Ltd., London, 1966.

ARTICLES

(for source see list of periodicals on p. 269)

17. "The pulse beat of maintenance today," (A), June 1966.
18. "Survey mirrors today's maintenance management," (B), March 1970.
19. "Your maintenance management—fit or flabby?," (A), January 1963.
20. "Eight fundamentals of preventive maintenance," (A), Vol. II, No. 8 pp. 87–90.
21. "In-plant communications: a survey of practices," (A), September 1962.
22. "Work measurement in maintenance," (A), January 1955.
23. "Work sampling in maintenance," (A), January 1959.

24. "Beating the breakdown rap," (A), December 1955.
25. "How to measure morale," D. G. Moore and R. K. Burns, (A), February 1956.
26. "Attitude survey uncovers employees' hidden discontent," (B), March 1970.
27. "Maintenance incentives hold the cost line," D. D. Swett, (B), September 1969.
28. "Maintenance planning/scheduling and controls," G. W. Smith, (D), February 1969.
29. "Maintenance organization: what's the best for you?," (C), December 1967.
30. "How to schedule inspection of electric motors," (A), June 1962.
31. "Industrial engineering organization and practices", Knut Holt, *The Journal of Industrial Engineering*, August 1968.
32. "Effective maintenance," D. F. Thomas, *The Journal of the Cambridge University Engineering Society*, pp. 77–88, 1955.
33. "The human element in maintenance," V. Z. Priel, (F), September 1967.
34. "The advantages of preventive maintenance," W. M. Copper, *The Journal of the institution of Production Engineers*, pp. 237–246, 1963.
35. "Management planning and preparation for work measurement," R. N. Barton, *Work Study*, June 1969.
36. "How to manage maintenance," J. J. Wilkinson, (the UMS system) (H) Mar–Apr 1968.
37. "One more time: how do you motivate employees?," F. Herzberg, (H) Jan–Feb 1968.
38. "Stimulation of effort," I. L. H. Scott, *The Production Engineer*, April 1970.
39. "Organizing executive time," Peter Drucker, *Management Today*, April 1967.
40. "Management ratios and interfirm comparison for management," *B.I.M. Management Information No. 3*, B.I.M., Management House, Parker Street, London, W.C.2.
41. "Maintenance performance—a decision problem," A. K. S. Jardine and W. Armitage, *The International Journal of Production Research*, Vol. 7, No. 1, 1969.
42. "*3-D Theory of Managerial Effectiveness*," copyrighted by Managerial Effectiveness Ltd., Box 1012, Fredericton, New Brunswick, Canada.
43. "Diagnostic documentation," J. B. Langham-Brown, Lt. Cdr., Royal Navy, (F), November and December 1972.
44. "How maintenance is measured," P. D. Guy, (the UMS system) (F), September 1972.
45. "Maintenance cost comparisons," D. Parkes, (F), August 1971.
46. "Computer improves plant performance," J. D. B. Steedman, (F), June 1973.
47. "Preventive maintenance guidelines," A. Kelly, (F), June 1972.
48. "Auditing maintenance," E. G. Freemont, (F) April 1972.
49. "Instant maintenance yardstick," V. Z. Priel, (B), July 1966.
50. "Twenty ways to track maintenance performance," V. Z. Priel, (A), March 1962.
51. "World-wide acclaim for maintenance," D. Parkes, (G), September 1972.
52. "Before you computerize—a consultant's advice," V. Z. Priel, (G), November 1973.
53. "Terotechnology, Concept and Practice," Department of Trade and Industry, London, 1973. (Available from Department of Trade and Industry, Industrial Technologies Secretariat—M4, Room 540, Abell House, John Islip Street, London, SW1P 4LN.)

PERIODICALS

A. *Factory Management and Maintenance* (later *Modern Manufacturing*), McGraw-Hill Inc., 330 West 42nd Street, New York, 10036.
B. *Modern Manufacturing*, McGraw-Hill Inc., New York, 10036.

C. *Mill and Factory*, Conover Mast Publishing, 205 E. 42nd Street, New York, 10017.
D. *Plant Engineering*, 1301 S. Grove Avenue, Barrington, Illinois, 60010.
E. *Maintenance Management*, Newsletter of the Maintenance Advisory Service, Silver Glade, Gasden Lane, Witley, Surrey, GU8 5QB.
F. *Maintenance Engineering*, Mercury House, Waterloo Rd., London, S.E.1.
G. *FACTORY*, Morgan-Grampian Ltd., Woolwich, London, S.E.18.
H. *Harvard Business Review*, Soldiers Field, Boston, Massachusetts.

FORMS AND SYSTEMS

Lamson Paragon Ltd., Paragon Works, London, E16 1NW.
Litton Business Systems, Brighton Rd, Sutton, Surrey.
George Anson & Co. Ltd., High Rd, Ilford, Essex.
Carter-Parrat (Visirecord) Ltd., Orchard Rd., Sutton, Surrey.
Kalamazoo Ltd., Northfield, Birmingham.
Rotadex Ltd., Kitts Green, Birmingham.
Farrington Data Processing Ltd., Telford Way, Acton, London W.3.

Index